Fire and Rescue Incident Command

A practical guide to incident ground management

Tony Prosser and Mark Taylor

FIRE
The trusted voice of fire & emergency since 1908

Pavilion

Fire and Rescue Incident Command
A practical guide to incident ground management

© Pavilion Publishing and Media

The authors have asserted their rights in accordance with the Copyright, Designs and Patents Act (1988) to be identified as the authors of this work.

Published by:
Pavilion Publishing and Media Ltd
Blue Sky Offices
Cecil Pashley Way
Shoreham by Sea
West Sussex
BN43 5FF
Tel: 01273 434 943
Fax: 01273 227 308
Email: info@pavpub.com

Published 2019

A catalogue record for this book is available from the British Library.

ISBN: 978-1-912755-09-7

Pavilion is the leading training and development provider and publisher in the health, social care and allied fields, providing a range of innovative training solutions underpinned by sound research and professional values. We aim to put our customers first, through excellent customer service and value.

Authors: Anthony Prosser and Mark Taylor
Production editor: Mike Benge, Pavilion Publishing and Media Ltd
Cover design: Emma Dawe, Pavilion Publishing and Media Ltd
Page layout and typesetting: Emma Dawe, Pavilion Publishing and Media Ltd
Printing: Severnprint Ltd

'There is an important distinction between command and effective leadership... Command is about authority, about an appointment to position – a set of orders granting title... Those who are privileged to be selected for command should approach those duties with a sense of reverence, trust and the willingness to sacrifice all, if necessary, for those you lead. You must love those you lead before you can become an effective leader. You can certainly command without that commitment, but you cannot lead without it: and without leadership, command is a hollow experience – a vacuum filled with mistrust and arrogance.'

General Eric K Shinseki, US Army, 2003

Contents

About the authors

Mark Taylor

MA, PG Cert HE, FHEA, GI Fire E

Mark worked for the Fire and Rescue Service for over 30 years in strategic roles including Operations Commander, Head of Command and Operational Training and Head of Terrorism and Contingency Planning for West Midlands Fire and Rescue Service. He has been involved in a wide range of major incidents and exercises at national and international level, including the 2007 floods and the Birmingham riots. As a senior lecturer at the Emergency Planning College and the Fire Service College, he has developed a range of programmes and courses which meet both practical and organisational needs of national and local Government, fire and rescue services, businesses and the academic requirements of students and tertiary institutions.

E-mail Mark at mark.taylor@artemistdl.co.uk

Tony Prosser

BSc(Hons), MSc, MBA, MEPS, FHEA, FI Fire E

Tony served in the UKFRS for 30 years including 23 years as a senior operational command officer in roles from Station Commander to Brigade Commander. His managerial roles include Director of Operations, Strategic Lead for Development and Head of Fire Protection and Prevention in West Midlands Fire Service and Head of Fire Protection in Oxfordshire FRS. He has been responsible for some of the largest incidents in West Midlands in recent years! He is a senior lecturer at the University of Wolverhampton and has worked as an incident command assessor at the Fire Service College, taught emergency planning and incident management for the Emergency Planning College and the International Fire Training centre. He has been a correspondent for FIRE Magazine for over 15 years and writes on operational, fire and community safety and FRS political issues.

E-mail Tony at tony.prosser@artemistdl.co.uk

Artemis Training and Development

On leaving the FRS in 2012, Mark and Tony formed Artemis Training and Development to deliver incident command training to a wide range of public and private sector fire and rescue services, airport rescue and firefighting services, national companies including National Grid, Network Rail, High Speed 1 and created a series of Fire and Rescue Degree programmes at the University of Wolverhampton www.wlv.ac.uk.

www.artemistdl.co.uk

Dedications

This book is dedicated to the memory of the firefighters who made the ultimate sacrifice in the service of others and the families and friends they left behind.

Mark

For Deborah and Arthur for being there whatever the weather.....

Tony

For Lesley, Eirian, Cerys and Marian with apologies for 30 years of missed birthdays, Christmas lunches and school plays and sports days......

Acknowledgements

We would like to thank the following people for their help and support for providing ideas, material and commentary of the material: Steve Bunt, Dave Martell, Andy Hall, Billy McGill, Justin Jones, Matt Elder, Steve Clack, Andy Huntley, Adrian Carter and the support of their services in Surrey, Hertfordshire, East and West Sussex Fire and Rescue Services. We would also like to thank the team and our friends at Artemis Training and Development Ltd, for their continuing support while the book has been in progress: Jim Sinnott, Sheetal Panchmatia, Lee Baker, Andy Kirk and Rory Campbell. Phil Hill and Nigel Adams have been invaluable in assessing the practicality of this book, and finally to Bill Gough MSc, QFSM, who remains, as always, a font of knowledge of all things operational. Finally, our thanks to Clare Williams, University of Wolverhampton, for indulging us while we have been working on this book.

We would also like to acknowledge Andrew Lynch, Editor of *FIRE* Magazine, who has been encouraging us since the beginning of the process and Mike Benge for his guidance and patience during the editorial process.

Any errors in the text remain ours and would ask to hear from readers who can provide corrections, enhance aspects of the book or provide illustrative incidents which may be of importance in subsequent editions.

tony.prosser@artemistdl.co.uk mark.taylor@artemistdl.co.uk

Chapter 1: Introduction

Whether the sound that wakes you in the small hours is the clanging of the station bells, a pager alert or a telephone call in the home, the noise sets off a whole chain of thoughts, expectations and options that race through your mind.

The role of an Incident Commander (IC) in the fire and rescue service is full of challenges, the like of which most people will never experience. When you arrive at an incident, a whole range of people expect you to arrive with all the answers to their problems, providing the central focus for a wide range of activities that will ultimately control the chaos of the incident. This will include members of the public, representatives of other organisations and, most immediately, those for whom you have direct managerial control.

The skills that the IC requires are wide-ranging and learned over many years by way of study: knowledge, skills, understanding, training, experience and practice, practice, practice. The image of an incident ground commander as portrayed on TV and in films as the all-knowing, all-seeing and wise leader is perhaps little more than a myth, because in reality we all need to work on these skills all the time, whether you are new in the role of leader or experienced in the widest range of incident types.

The overall aim of this book is therefore to guide you through some of the key issues that face ICs in the modern fire and rescue service, to identify good practice where possible ('best practice' is a term always open to challenge), to identify examples of real incidents where lessons have been learned the hard way, and hopefully to set challenges for you to think about in your most important workplace: the incident ground. What this book cannot teach is the face-to-face skills that are required of all ICs, which you will need to learn and practise on the training ground, at incidents and exercises.

The development of incident command systems in the UK

While organised firefighting has been present in the British Isles since the invasion of the Romans in 43 A.D., few organised systems of incident management were been developed and an ad hoc approach was taken to controlling fire, usually undertaken by the military, when available, or by local citizens working together to save property – saving lives was considered relatively unimportant until the 19th century.

In the early hours of the second of September 1666, a fire broke out in Pudding Lane, London. As the fire spread, ineffective measures were undertaken to control the conflagration, which finally led to the Duke of York being appointed – over the head of the Lord Mayor of London – to control the fire. After four days the fire was surrounded, and through the use of fire breaks, created by naval dock workers and other military personnel using gunpowder, eventually it was controlled. One witness said that, *'had that the Duke not been present, and forced all people submit to his command, I'm confident they had not been a house standing'*. What the Grand Old Duke of York had done was to take charge of the incident, organise the incident ground, direct resources effectively to control the outbreak, ensure the safety of the community and preserve property as best as possible. Any IC would recognise these key tasks and activities as those which they themselves undertake at every incident they attend.

While London started to develop its fire services, as did other large cities (generally on insurance service basis), scenes at incidents were sometimes as chaotic as that of the Great Fire of London, but gradually, with organisation, firefighting became more structured, especially once serious consideration was given to the business of firefighting. James Braidwood, the Scotsman who founded the first municipal fire brigade in Edinburgh in the 1820s and the first superintendent of the London Fire Brigade establishment in 1833, was one of the great thinkers and innovators of the fire service in Britain. In his book *Fire Prevention* and *Fire Extinction*, published after his death in 1861, he succinctly explained the essence of incident command (emphasis added):

> *'It is now generally admitted that the whole **force brought together** to extinguish a fire ought to be under the **direction and control of one individual**. By this means, all quarrelling among the firemen about the supply of water, the interest of particular insurance companies, and other matters of detail, is avoided. By having the whole force under the command of one person, **he is enabled to form one general plan of operations**, to which the whole body is subservient; and although he may not, in the hurry of the moment, at all times adopt what will afterwards appear to be the best plan, yet it is better to have some general arrangement, than to allow the firemen of each engine to work according to their own fancy, and that, too, very often in utter disregard as to whether their exertions may aid or retard those of their neighbours.'*

He also emphasised the point that this continuity of command (the chain of command) should not be usurped by others:

> *'I need scarcely add, that on no account whatever should directions be given to the firemen by any other individual while the superintendent of brigade is present; and that there may be no quarrelling about superiority, the men should be aware on whom the command is to devolve in his absence.'*

These principles still hold true today and form part of the basis of effective incident command, which basically enables one commander at an incident to control, co-ordinate and command others to ensure that a plan is developed, implemented, monitored and amended. This can be achieved through a number of methods including direct command, delegation and setting up structures to support incident resolution, but always maintaining the focus and direction set by the IC.

Figure 1.1: James Braidwood

The greatest challenge to incident command and control before the 1990s was, somewhat obviously, the large-scale attacks on the UK from the air during the Second World War. Unprecedented numbers of simultaneous fires, and the scale of these fires, necessitated the mobilisation of sometimes several thousand firefighting appliances and tens of thousands of firefighters. During the largest raid on London (the night of 29th to December 30th, 1940), over 2,000 pumps were at work fighting 1,500 fires. Fourteen firefighters were killed that night, and over 250 seriously injured. To meet the challenge of these large fires, command and control procedures were developed that had not been used at 'routine' pre-war incidents. The predominant tactics were to contain the spread of major fires, preventing them joining up and creating conflagrations – uncontrolled fires sometimes called 'fire storms' where the thermal energy and updrafts are so great that they create their own winds that fan the flames. The consequences of not managing these major fires were seen at Hamburg and Dresden later in the war.

Following the war, Britain's fire services settled back into a relatively routine existence, although the number of services was reduced from the pre-war 1,502 to around 150.

Thanks to the standardised equipment that had emerged during the war, similar procedures for dealing with large fires and other incidents were also developed. Despite this, however, incident command control at some larger incidents remained problematic.

The deaths of firefighters at fires in Covent Garden (1949) and Smithfield Market (1958), London, gave rise to challenges to existing procedures for maintaining knowledge of those attending incidents and their deployment (sometimes called operational accountability) and the use of breathing apparatus. Eventually these deaths led to more detailed control of firefighters on the incident ground and set out procedures for the use of breathing apparatus. As far as incident command was concerned, until this point guidance had hardly evolved since the war despite the introduction of new equipment, resources and tactics.

The development of a systematic approach to incident command

By the early 1990s it was becoming increasingly apparent that fire ground organisation, accountability systems and command skills were not developing in pace with the increasing complexity of incidents being attended. Increasing enforcement action in the form of Improvement Notices by the Health and Safety Executive on the fire and rescue service, notably following several incidents where firefighters had died – Hayes Business Services, Gillender St, London, 1991, Sun Valley poultry processing plant in Hereford in 1993 and others – led to concerns being raised by several in the fire and rescue service that incident command was not being implemented efficiently or effectively and that a more systematic approach to controlling complex incidents was necessary.

One of the pioneers to develop an incident command system (ICS) in the UK was Kevin Arbuthnot, then Assistant Chief Fire Officer in West Yorkshire Fire and Rescue Service. Kevin developed a three-strand approach to incident command. Taking the processes developed by Phoenix Fire Department, Arizona (then led by Alan Brunacini), he considered issues including the assessment of operational risk at incidents, organising the fire or incident ground, and finally, the competence of incident commanders themselves. In 1999, this ultimately led to the development of a ground-breaking approach to incident command (at least in the UK) – having a fire service manual, which for the first time gave the whole of the UK a standardised approach to the management of incidents.

Although some were less than pleased with the introduction of some American terminology – 'offensive' and 'defensive' tactics – it was widely accepted that the manual filled a significant gap in the way incidents would be managed in the future. Several iterations of incident command manual have since been published, and following the introduction of the National Operational Guidance

(NOG) programme, a new document which serves as 'the foundation for incident command' has been issued and covers some of the material originally produced for the fire service manuals.

It is these NOG documents that this book is intended to supplement and add some 'colour' to flesh out the content and deal with the principles of incident command, and not all of the precise details in the NOG documents which are regularly amended.

How to use this book

This book aims to add to existing materials on incident command and to provide current and potential ICs with a broad understanding of what is expected of them and the processes and systems that the UK fire and rescue service uses to safely command an incident. Current NOG documentation runs into many thousands of pages and it may be daunting to attempt to read all the materials available. We have therefore aimed to summarise and condense the essence of this guidance into a single practical volume – this is not intended to be used as an abstract, theoretical tome, but is founded on decades of operational experience on the incident ground and working with many experienced fire officers, too many to name, who have been kind enough to share with us their experiences and wisdom. While the book focuses on guidance produced mainly for the UK (and sometimes only parts of the UK) fire and rescue services, we have also sound principles from other parts of the world which seem to have value for all firefighters.

We have included in the book examples of a wide range of incidents and the lessons that can hopefully be learned from them. These examples are either documented in national or local archives or through the personal experience and direct interviews with individuals who attended them. Due to the sensitivity of organisations to criticism, we have decided to omit details of services and individuals, although many are readily identifiable. Some of these incidents were controversial and remain so, but it is felt that openly recognising the key failings of others (and ourselves) aids rather than hinders learning.

Despite the fact that today, in 2019, we are fortunate to have seen no firefighter deaths in the UK for over five years, there are still too many close calls to enable us to say that we have sufficiently learned the lessons of the past. We hope that even if this book is only dipped into occasionally, some of the ideas will jog your memory at the next incident you command.

Chapter 2: Leadership and command at critical incidents – an immediate challenge

Introduction

At all operational incidents, the IC will be expected to lead, monitor and support the activities of their teams who are physically resolving the incident. The extent to which they will be doing each of the those activities will depend very much on the type of incident, the number and lines of supervision that are needed to control the tempo and dynamic of the incident (sometimes called the 'battle rhythm', and defined as a deliberate and controlled cycle of command, control and operational activities intended to synchronise current and future operations) and the relative experience of those under the command. The command of an incident for the first time can be daunting but a systematic approach, no matter the size of the event, will help the IC resolve the incident in a controlled fashion.

These systems, particularly when combined with a decision support tool such as the DCP (Decision Control Process), when enhanced by experience across a range of events of different sizes and complexities, allow the commander to grow in confidence and develop the ability to manage large numbers of staff and equipment to resolve even the largest of incidents. This chapter will consider the personal skill sets that an IC requires, the separate components that make up the IC's responsibilities, and how the levels of command are graduated to support the resolution of the widest range of incidents. These skills include both the technical command skills – knowledge of systems, processes and protocols – and the behavioural skills required to make sure plans are carried out effectively.

An individual's ability to perform the role of IC will have been identified through a selection process, a significant part of the safe person concept. Such selection processes will invariably include a technical assessment of the person's acquired knowledge and a practical demonstration of their command skills in a practical or simulated incident. Having demonstrated this potential, the individual will then undertake a period of 'upskilling' and development, followed by a graduated process of gaining practical experience of command, firstly under supervision and/or coaching by a more experienced commander, and then, after being deemed competent, allowed to command an incident in its totality. It is not the intention of this book to set out assessment and selection criteria for command since there is currently no set process in the UK.

Technical command skills

To manage an incident effectively, the commander has to take account of many often competing factors, and develop a plan of action that will neutralise the hazards and protect life and property, while ensuring the risks that firefighters are exposed to are minimised. To achieve this the commander will need to:

- adopt systems and measures to ensure that command and control of the incident is maintained at all times

- adopt standard processes to ensure that hazards and risks are identified, assessed and mitigated as part of a continuing activity

- ensure that resources are adequate for current activities as well as sufficient to deal with any extra demands caused by unplanned events or reliefs

- develop a systematic approach to producing a realistic and prioritised incident management plan using the resources in attendance or en route

- manage the allocation, deployment and co-ordination of resources to meet the demands of the incident through sectors

- communicate the plan effectively to those who are charged with carrying it out and establish radio networks for monitoring, co-ordinating and controlling activity, supporting teams and sectors

- assess the success of the plan through constant evaluations of progress against planning assumptions and expectations.

The technical skills and knowledge associated with incident command are addressed in detail in chapters 6–12.

Underpinning knowledge and understanding

It goes without saying that the competent IC will have a good working knowledge of systems, information sources and technical skills to enable them to command an incident effectively. The problem facing us all today is the vast amount of information that is available and the limited amount of time any reasonable person has to read, digest and learn the salient points. One source of guidance in the UK has around 8,000 documents available, together with list of identified hazards and tactical and strategic actions that may be required. If each page takes just three minutes to read, over 600 hours are required to peruse (never mind understand) this body of guidance. And then it is only guaranteed to be correct at the time of downloading and printing! Nonetheless, it is still incumbent on the IC to have a detailed knowledge of matters pertinent to their current role in the service and at the incident ground.

As far as incident command is concerned, there is a continuum along which the level of technical knowledge and skill required varies in proportion to the seniority of the role being undertaken. A firefighter riding a fire engine, for example, will have a great deal of technical knowledge and the skills associated with the equipment they are required to deploy in their current role. They will need to know about the use of pumps, breathing apparatus and hydraulic rescue gear in great detail. However, they will not necessarily need to know about incident command systems in such detail but they will need to know how their role fits into the ICS, have an overview of the ICS system, and know how to undertake the role of a command point operator.

If they are a command support unit (CSU) operator, a detailed knowledge of equipment, systems, procedures and principles of ICS is essential, as are the technical skills of managing the IT and communication systems on a CSU. Supervisory ICs (Level 1) are required to have knowledge of ICS systems because they will be the first commanders on scene and have a duty to implement ICS as part of a safe system of work. They will still be required to supervise and sometimes operate equipment, and also carry out supervisory command roles such as Level 1 IC, as well as supervise/operate other safe systems of work including sector command and breathing apparatus entry control.

With increased levels of responsibility for the management of larger and more complex incidents, the requirement for a more detailed understanding of the incident command system and its application becomes greater, while the demand for knowledge of the 'hands on' practical tasks diminish. At level 2 incident command significant incidents could be managed by a middle manager who will have a greater understanding and appreciation of the incident command system, including a detailed knowledge of command support operations and other supporting functions at an incident.

With increased responsibility there is a reduced requirement to deal with tasks on the incident ground, and commanders must remember to 'command not act', in the words of Alan Brunacini. Commanders should become increasingly directive and able to delegate effectively to other officers. The figure below shows the relationship between skills, knowledge and understanding for command skills and core operational skills

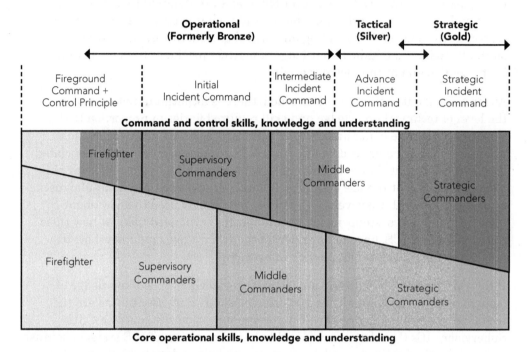

Figure 2.1: Knowledge, technical skills and incident command skill requirements

IC leadership behaviours and skills

The skill sets and behaviours of the IC are as varied as the definitions of leadership and leadership traits. Every now and then a new set of behaviours is identified and is sometimes adopted by a service, or even a national body. There are many common skills listed in hundreds of lists of leadership traits and behaviours that have been derived from a wide range of sources. Some are particularly applicable to command and leadership skills at critical incidents involving the FRS. While the following skills are suggested for consideration, it is important to recognise that leadership and its associated behaviours remain a matter of debate and we have based the following on suggestions made by studies through the National Operational Guidance Programme (2018), John Adair (2002), Flin, O' Connor and Crichton (2008), Alan Brunacini (1985) and The Home Office (Fire Departments) (1981).

The ability to make decisions

The ability to make decisions during a critical period is a key skill of the commander. Knowledge of decision-making styles and processes is important, but the *willingness* to make a decision, even with less than a complete knowledge, is vital. To paraphrase a much misquoted truism: *'The best decision is the right decision. The next best decision is the wrong decision. The worse decision is no decision'* (Scott McNeally). You can correct a bad decision, but to make no decision due to procrastination – paralysis by over analysis – may result in firefighters or the public being put at risk for extended periods while waiting for the fire (or other hazards) to make the decision for you. Good ICs will lead the way, using their technical knowledge and professional judgement to make critical decisions in the absence of the full facts, balancing potential risks of harm against possible benefits.

Firefighting is a risky business, but indecisiveness allows risk to outweigh potential benefits.

Assurance and confidence, both in themselves and in their firefighters

An important underpinning behaviour of the commander is the display of confidence, not only in themselves and their abilities but also in those who are working as part of their team. Nothing demonstrates self-confidence better than displaying faith and belief in those under your command who have been given responsibility for achieving objectives or tasks unsupervised or with minimum oversight. The incident ground can be a busy place and the IC who is constantly interfering with a subordinate's activities rather than simply monitoring them, may undermine the confidence of the individual and slow down processes, potentially delaying incident resolution. It is also true that not many ICs can possibly have a full range of technical skills and knowledge to manage a large or complex incident alone. Most of us have a good working knowledge of many things we need to know. What we need is the ability to outsource those skills and the knowledge we don't possess by having a good understanding of the resources that a FRS can access. The IC should know what is available, where it can be accessed and what skills come with the firefighters under their command. Apart from the tactical advisers and subject matter experts with specialist skills in Hazmats, fire investigation, water management etc., there may also be firefighters with specialist knowledge from a previous career (or current career in the case of 'on call' or volunteer firefighters), all of which adds to the arsenal of knowledge that can be deployed at an incident.

Calmness

In the UK in the 20th century, it was deemed that 'officers' (read 'ICs') should, *'appear to be at all times imperturbable, and any orders should be given clearly and simply, without shouting or signs of excitement'.* Notwithstanding that the terminology has changed in the intervening years, the ability to remain calm during a moment of crisis is invaluable. The commander who turns up at a well-developed fire and races around the incident ground with tunic unzipped, helmet missing, shouting at crews and in a panic is hardly likely to inspire confidence in firefighters. The commander who arrives at the incident, dresses correctly while listening to the fireground radio, assesses the 'mood' of the incident ground, observes how the firefighter teams are behaving and who walks to the Command Unit or command point will help to calm concerns and provide a focus for the leadership of the incident, making the incident 'business as usual' and not a drama or a crisis. At critical times, a commander who remains calm and focused will be able to organise and decide actions more effectively than one who contributes to an atmosphere of increasing chaos.

Willingness to accept responsibility

All personnel at an incident should take personal responsibility for the way they act and behave to ensure their own, their colleagues', and others' safety. Although this may seem an obvious behavioural trait and/or moral expectation of a leader, it is important for your team members to know that you are ultimately responsible for their safety, the management of the incident and the decisions taken on your behalf. With authority goes responsibility and the most senior commander on the incident ground, irrespective of whether they have taken the role of IC or not, holds legal, corporate and moral accountability for the incident. Clearly not all commanders will necessarily have the depth of knowledge of an experienced commander, especially in their early years, but providing the support and shouldering blame for when things go wrong sets a high standard of behaviour that engenders respect and trust for decisions that will bear fruit as incident ground relationships develop with time.

Personal integrity and honesty is a trait that is often seen by most as essential for a leader, and nowhere is this more important than on the incident ground where a commander uses this authority to put firefighters in harm's way. It is equally important that the commander understands the limits of their authority, particularly with respect to operating within a multi-agency context, where firefighter safety is concerned, and where individual responsibilities for safety outweigh those of the organisation.

Determination

Once a plan is set in place, the IC may sometimes need to show a determination to 'see things through to the end', and to be resilient in the face of criticism or reluctance to implement orders. This will need a clarity of thought to balance the needs of the incident, the needs of firefighters and the risks being faced. Sometimes the resistance may be operationally based ('That won't work – it didn't last time!') and sometimes political ('You can't leave that barn burning – it has pesticide in it!'). Confidence and conviction, an empathetic ear and a clear explanation of the task and reasons for the (well thought out) plan will help convince sceptics and show your determination to complete the task successfully.

Very often however, things won't go as expected, exemplifying Von Moltke's dictum that *'No plan survives first contact with the enemy'*. At this stage, the commander will need to show an ability to evaluate a plan critically, modify where necessary and set in motion actions to resolve the incident. Accepting that Plan A has failed and that Plan B is required will demonstrate the ability to be flexible with changing circumstances. Determination is one thing, but there also comes a point where one needs to recognise that a plan needs to change and that new tactics are needed.

Enthusiasm

Most firefighters take on a career in the FRS to serve the community in a meaningful way. There is no more visible service than attending an operational incident, and at most incidents (at least in the active phases) firefighters of all roles and positions relish the opportunity to use those skills they have been training in for years. A good commander will tap into this enthusiasm and express his willingness to lead the incident, using good will to achieve rapid success. It is not a failing to demonstrate your skills and ability to command. It is, however, important that in your enthusiasm you do not take on roles that are not within the role of the IC – remember: you lead, monitor and support, you don't DO! A Level 4 IC who is directing jets inside a building is undertaking a task that a crew manager or sector commander should be directing. If the IC is involving himself/herself with task-level activities, it needs to be asked who is running the incident?

Box 2.1: Indecisiveness and unco-ordinated actions

A large fire occurred on a hot summer's day in a ground floor city centre shop that was part of a larger commercial block. Breathing apparatus (BA) crews began interior firefighting operations, repeatedly entering the shop and withdrawing when fire conditions worsened. Over a period of several hours the process of offensive operations and withdrawal continued. The unco-ordinated use of an aerial monitor at the front of the building is believed to be the cause of an increase in temperature at the rear of the building and it is believed that the worsening of conditions made a significant contribution to the death of a firefighter and the collapse of another.

Lessons learned

If a fire in a building continues to burn for several hours, following the same tactics that have already been employed for the last three hours is not the sign of a commander showing determination. It is rather an indicator of stubbornness or moribund thinking, and indicates the lack of ability to generate more creative options for dealing with the incident. Start thinking creatively!

A lack of effective control and an unco-ordinated approach to tactics employed on the incident ground can contribute to an increased risk for firefighters and bring about a deterioration of incident conditions.

People skills

It is often forgotten that commanders for the most part are dealing with individuals and not with objects. It is imperative that commanders at all levels actively demonstrate respect for all staff on the fireground both in word and deed. Furthermore, this common decency must be extended to the other agencies and staff in attendance, and especially to the members of the community, who are the people who pay the wages of firefighters of all ranks and deserve all of the courtesy and good manners possible. This empathy and sympathy for your colleagues and others is a hallmark of a good leader and puts the whole service in a good light, an important factor in maintaining public and staff support away from the incident ground.

Despite dealing with the difficulties in trying situations, it is important for the commander to maintain a balanced sense of proportion and a positive mind set. This includes not being afraid to show that you are human and, if the circumstances are appropriate, show you have sense of humour. A calm, confident and approachable IC will be able to achieve far more than one who issues orders and instructions without taking time to encourage and enthuse those who have to carry out the task. The more challenging the task, the more time should be spent on briefing those who have to carry it out.

Communication skills

The ability to communicate your intentions succinctly and efficiently is key to getting tasks achieved as effectively as possible. Having a grand scheme will fail if it is not understood by the firefighters who have to carry the instructions out. There are numerous tools and techniques that are available for briefing teams and individuals such as the NATO briefing tool SMEAC, which is a particularly useful device for briefing task-level activities:

- **Situation.** Give a basic account of your current situation.

- **Mission.** Give a one-line statement of the objective of the mission (then repeat the statement).

- **Execution.** This is basically your plan for the mission, which you explain in as much detail as possible to the team. You could use visual aids such as diagrams.

- **Any questions.** Simple enough – ask your team if they have any questions, but watch your time.

- **Check understanding.** Question your team on the plan – who is the Sector Commander or safety Sector Commander? What are the limitations and the time restrictions?

For more complex issues and incidents involving larger numbers of staff and equipment, more detailed briefings may be required that necessitate the production of plans, diagrams and written instructions. While the temptation may be to rush such briefings, it is important to remember that the task/activity briefing is a key part in providing a safe system of work at incidents and needs to be as full and complete as available information will allow. A good IC will seek to develop a dialogue with their team at an incident by effective two-way communications – this will have the twin benefit of improving engagement with the team and also encourage the development of ideas so that outcomes are improved through the use of ideas that have not just come from one source. The issue of incident ground communications is looked at in more detail in Chapter 12.

Dealing with conflict

Although the incident ground should be a positive working environment where a culture of mutual trust and respect exist, sometimes, in stressful and pressurised conditions such as those that can exist at incidents and in the command unit, it is unsurprising that conflicts can arise. These may include disagreements about operational priorities and tactics, the organisational structure, the speed at which progress is being made and the roles being allocated to staff. As IC and the leader of the team, handling this conflict is a key skill and is necessary to ensure that relationships are productive and positive at all times. Where conflict occurs there are a number of things that can be done to help reduce tension and improve relations between team members. But it is important to emphasise that action must be taken rapidly to avoid compromising the IC's operational plan and firefighter safety.

Awareness of the behaviour of others and how they interact with colleagues may help alert you to an impending disagreement or argument. Sometimes the signs may be obvious, such as raised voices or changed tone and arguing. At other times an individual may express antagonism or disengagement by their body language, such as folding arms, avoiding eye contact or negative facial expressions – raising eyebrows or frowning, for example. When you notice such signs, it may be possible to take action to prevent escalation of the disagreement and calm everyone down.

Where signs of conflict have been spotted it is important that action is taken as soon as possible to remove the cause of the conflict. Taking a neutral position, the IC should be impartial and seek to get to the bottom of the problem. Where there is a difference of opinion on a technical matter, or contradictory orders or instructions, the IC should seek to gather both views and arbitrate on the resolution quickly. Where the issue is a conflict of personality then it may be appropriate to make alternative arrangements to create physical space between the two sides. For the sake of operational expediency, this may necessitate one member changing roles to avoid contact between the two.

Whatever the cause of the conflict, it is essential that the matter is not closed with the end of the incident. Failure to resolve a conflict between individuals can lead to future disagreements and have an impact on operational efficiency and the organisation. ICs should seek to resolve the conflict outside the operational environment and seek appropriate advice to ensure that differences can be reconciled in a positive manner.

Personal skills improvement

A professional IC will have most or all of the skills listed above to varying degrees. The more self-aware the commander is of their skills, abilities and knowledge gaps, the more likely they are to be able to plug those gaps. In order to develop those skills, there are many tools available to analyse a wide range of personal traits and skills. As part of a personal development plan, it is often useful to carry out a 360° assessment of skills using operational feedback from peers, line managers and those being managed. This can help identify areas that need working on and can provide reassurance where strengths are recognised. In some FRSs, operational performance is used as part of annual appraisal procedures which helps develop the individual and improve the service's response to incidents. Another method of appraising self-performance is to develop a network of peers to provide post-incident performance feedback, coaching and mentoring.

Summary

You should now understand the following terms, ideas and concepts explored in this chapter:

The technical skills associated with commanding incidents – an understanding of the standard operating procedures, systems and technology used by your organization to manage critical incidents, and how this changes as the command levels increase

The personal behaviours and qualities you should develop and use to command incidents – these include decisiveness, self-awareness and self-confidence, remaining calm and determined in the face of challenges, accepting personal accountability for your actions, and people skills to get the most out of your teams.

You should be considering how you will assess your skills, identify gaps in your skill set and create a development plan and actions to plug those gaps.

Chapter 3: Fireground strategy, tactics and tasks – a refresher

The aim of this chapter is to provide a brief revision of strategy, tactics and task-level activities to control an incident and make an effective start down the road to recovery. It considers the essentials of the three levels of incident activity and explores some of the issues that may be pertinent for the IC. It also considers the sequence of events involved in the control of an incident, in both generic and specific contexts, for the more common types of incident.

Introduction

While it is strictly true that all incidents are different, there is also a great deal of commonality between them. This means that we can develop systems of work that allow us to make a relatively quick assessment of what is happening and how it may be addressed through a relatively standardised process of risk analysis and assessment. The development of an 'all hazards' approach allows firefighters to use a systematic command process for the resolution for all incident types, whether fires, transport incidents, hazardous materials spillages etc. Nevertheless, despite this systematic approach, the differences between what we think *should be* happening and *what is* actually happening in real incidents provides a challenge and requires a greater level of analysis.

This is the difference between developing tactics based upon the almost wholly intuition-based and rapid 'recognition primed decision making' (RPD) and the slower and more considered analytical decision making. Fire and rescue services recognise that many incident types can use a standard plan of work – Standard Operations Procedures (SOPs), Tactical Operations Plans (TOPs) etc. – which will detail the steps to be taken to resolve an incident. The plans, often based upon national or international guidance, will invariably be customised to meet local requirements such as area risk, resource levels, equipment available and staffing models. This chapter will attempt to consolidate the various approaches to incident command that can be applied generically to most incidents.

Defining 'strategy', 'tactics' and 'tasks'

For the purposes of this chapter, the three terms – 'strategy', 'tactics' and 'tasks' – will refer to operations on the fireground and should not be confused with strategic, tactical and operational command levels (also known as Gold, Silver and Bronze among many organisations in the UK and elsewhere), which form part of the overall management structure for major disasters and emergencies.

There has been much debate within services both in the UK and overseas as to which activities on the fire ground constitute strategic or tactical or task level 'jobs'. The definitions offered here are a composite of the latest thinking in the UK and in North America.

Fireground strategy

At its most basic, strategy is the setting out of what is to be achieved at an incident by the fire and rescue services and other agencies. At this simple level, the strategic objective may be to contain a fire to its location of origin and prevent its spread to other locations.

At larger and more complex events, and even dispersed at multiple sites, the IC may need to consider a set of wider implications beyond the immediate incident before determining a strategy. At these incidents, where more than one agency may be involved, a more sophisticated method of co-ordinating strategy and tactics may be required. It is for this reason that the Joint Emergency Services Interoperability Programme (JESIP) has been developed in the UK, while other similar systems are used elsewhere: a co-ordinated system designed to reduce harm to people and the community.

Setting operational strategy and objectives

In order to determine the strategy for resolving an incident, the IC, whether attending a refuse bin fire with one 'pump' (a fire engine, appliance, etc), or a major incident, needs to be aware of the simultaneous and sometimes competing objectives they may face. These may include, for example, the need:

- to carry out rescues
- to contain the fire (or other hazard) in order to allow rescues to take place
- to take offensive activities to extinguish the fire (or neutralise the hazard) in order to reduce the risk to life
- to extinguish fire

- to protect the environment

- to minimise the impact of the incident on the wider community.

Incident strategy will be to set these objectives in order of importance: these are the IC's strategic objectives.

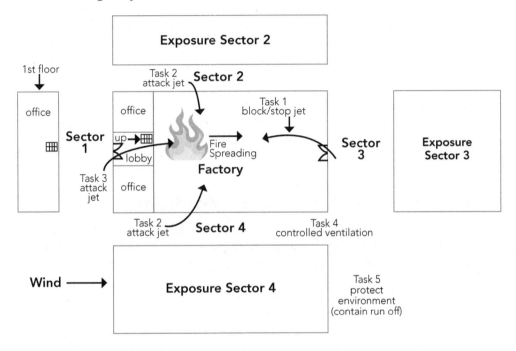

Figure 3.2: Schematic of operational incident priorities (drawing of a factory fire)

An example of an incident of greater complexity might be a fire in a farm building that involves pesticides, herbicides or other agricultural chemicals. In this incident, the IC would need to think about issues including the impact of any firefighting water on watercourses or rivers, the likely impact of a contaminated smoke plume in residential areas as well as the potential for spread to other buildings. It is possible that in this example, the strategy employed by the fire and rescue service may be to allow the farm building to burn while ensuring the fire does not spread to adjacent properties.

As mentioned, an issue to bear in mind is that other agencies and authorities may also have conflicting priorities. In this example, one agency may wish the fire to be put out because of the airborne risk of smoke to the community while other agencies may want to reduce the pesticide risk by allowing any contents of the building to be burned off. The IC will need to weigh the options available and seek consensus as to the best solution available when all factors have been taken into account and a risk versus benefit assessment has been carried out.

Fortunately, like the fire and rescue service, other agencies have begun to take note of wider risks and responsibilities, and in the UK, the Environment Agency (EA) and equivalent agencies in Scotland, Wales and Northern Ireland have developed sophisticated liaison and working arrangements with the emergency services. The use of environmental risk assessments (provided by the EA) and standardised action plans have enabled fire and rescue services to work together to reduce harm to the environment.

There are sometimes other considerations beyond the incident ground which may have an influence on the strategy eventually selected. Box 3.1 gives one such example, of how policy at a national level had a direct impact on the operational strategy at an incident itself.

Box 3.1: Fire and explosion in a fuel storage facility

In 2005, a major fire and explosion occurred in a tank form at the Buncefield oil storage terminal in the English Home Counties. An unconfined vapour cloud explosion was caused by a leak of petroleum and caused damage and set fire to over 20 large storage tanks. The initial tactics were determined to be a major foam attack on the day of the explosion. Concerns about the potential impact of pollution on local rivers and water sources delayed the attack by foam cannons using high-volume water pumping units, and consideration was given to allowing the fuel to burn off. However, the impact of the fire on the local area as well as the negative media perception of the incident (i.e. the fire was continuing to burn) and the smoke plume that was descending and reaching ground in Europe led to the decision that the fire should be attacked with large quantities of foam. The fire was rapidly extinguished using firefighting foam containing perfluoro-octane sulfonate (PFOS), the importation of which was about to be made illegal in the UK. Spillage of PFOS led to contaminated water (a holding tank was damaged during the initial explosion and leaked), which in turn caused contamination of limestone aquifers which provided drinking water for the region.

Lessons learned

- Operational considerations can be overridden by political/national considerations.
- Ensure that all potential consequences of a strategy are considered and discussed before implementation.

Similar considerations need to be taken at incidents involving hazardous materials. Strategic options may include:

- Taking action to contain the incident directly and carry out rescues through the use of firefighters and equipment within the hazardous area (also termed the 'hot zone' – see below).

- Containing the incident to a defined area by setting up appropriate cordons and awaiting the attendance of specialist environmental hazard waste teams.

- Where no immediate risk or need for intervention has been identified, providing advice to support the 'owner' of the problem in resolving the incident without the deployment of fire and rescue service assets.

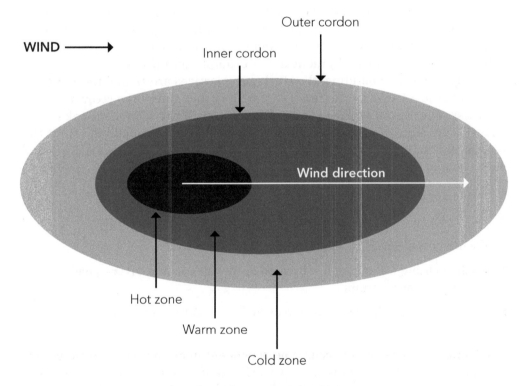

Figure 3.3: Incident zoning for a hazardous material incident

In many respects, dealing with road traffic collisions provides limited strategic options but essentially follows the same methodical process. Where life is involved, active intervention is necessary and should commence as soon as possible. Where life is extinct, the responsibility for the incident management is likely to belong to other agencies, notably the police service in the event of a serious transport accident. In this case, the strategy is likely to be to stand by and provide support to other agencies if and when required.

Following analysis of the incident (including the strategic objectives, the type and size of incident, the speed of development, other risks, and, fundamentally, the risk to safety of firefighters at the incident), the IC will determine whether the best resolution of the incident will be arrived at by either using offensive tactics (within the hazard zone) or defensive tactics (outside the hazard zone). A more detailed discussion on offensive and defensive operations can be found in Chapter 8, but we will now consider the implementation of the tactics required to control the incident.

Operational tactics and priorities

If strategy sets out 'what' is to be achieved, then tactics form the 'how' – the plan by which the strategy is going to be achieved. On the fire ground, there is a degree of overlap between strategy and tactics, and to a certain extent this is both understandable and practical. For example, at a fire where persons are trapped within a building, the strategic objectives are straightforward: carry out rescues, control the fire, extinguish it, and undertake recovery activities. The overall strategy and tactical plan will therefore be to take offensive actions, i.e. place firefighters within the hazard zone to carry out the rescues and extinguish the fire etc. The tactics required to achieve a successful outcome may include:

- the deployment of ladders

- the deployment of main jets by crews wearing breathing apparatus

- the use of ventilation in an aggressive format

- supplementing water supplies from an open water source (rivers, ponds, streams etc) or hydrants

- setting up a casualty clearing area or first aid point with the assistance of paramedics.

The tactics employed will of course be dependent upon the type of incident and the stage to which it has developed. Clearly, an attendance at a smouldering bin fire will require different tactics than a fully involved compartment fire. Similarly, the limited spillage of a chemical in a warehouse will be treated differently to an incident in which a person has been contaminated by that same chemical.

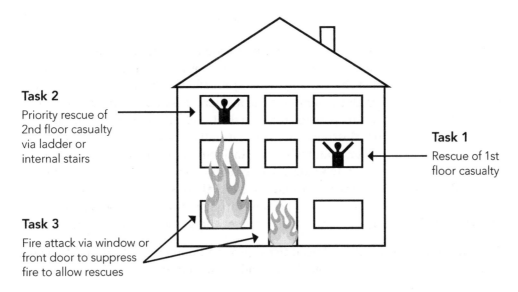

Task 2

Priority rescue of 2nd floor casualty via ladder or internal stairs

Task 1

Rescue of 1st floor casualty

Task 3

Fire attack via window or front door to suppress fire to allow rescues

Figure 3.4: Building fire

In Figure 3.4, a fire on the ground floor of a building has trapped people on the first and second floors. The IC must decide the priority for operational tactics. If the first attendance is a single pump with a crew of four (the IC and three firefighters), then tasks should be prioritised on carrying out rescues of those most at risk. In the case in Figure 3.4 the task priorities should be:

1. Rescue the person on the first floor as they are directly above the fire and at most risk.

2. Rescue the person on the second floor, who, although above the ground and trapped, is not directly above the fire and so at less immediate risk.

3. Attack the fire once casualties have been rescued.

When additional resources arrive, a simultaneous attack on the fire can commence and support rescue operations.

Incident ground tactics are the foundation stone of a successful operational skills outcome. Firefighters train and practise basic elements of incident procedures such as pitching ladders, deployment of breathing apparatus teams, the use of hoses and branches, chemical protection suits, hydraulic rescue gear as well as a host of more specialist procedures including the use of command units, aerial ladder platforms and rescue vehicles.

Tactics are to strategy what individual instruments are an orchestra: to make the music all parts must work together in unison.

Actions of the first and subsequent ICs

As Braidwood wrote, in order to ensure that incidents are managed effectively and to ensure that the intended outcomes are achieved, it is necessary that someone is effectively in control of the strategy, setting out the tactical parameters of the incident and for managing the tempo of the incident. Stretching the musical analogy a little further, this person is the conductor of the orchestra – on the incident ground, the IC. Most incidents are normally attended and dealt with by a relatively junior IC or, if and when the incident grows in size or complexity, more senior officers are likely to attend and take command, along with a range of specialist support officers and resources. Nevertheless, the principles that underpin incident management are present throughout its duration and need to be reviewed continually to ensure that strategy and tactics remain appropriate.

From first attendance, ICs will need to carry out a number of essential actions to support the successful resolution of the incident, including:

- Information gathering.
- An assessment of the incident or situation – deciding what the problem is.
- Assessing the risks.
- Determining the strategic objectives and tactical priorities.
- Identifying tactical options.
- Selecting one of the identified options.
- Setting an outline of tactics to match the needs of the strategic and tactical priorities.
- Assessing the resources required, based on those priorities.
- Implementing the tactics to control, contain and neutralise the hazards.
- Organising the incident: structuring the incident, sectors, sector tasks, allocation of resources, communications structure, administration (safety reporting, messages, reliefs, briefings, reporting schedules and updates/sitreps).
- Taking steps to mitigate the impact of the incident.
- Closing down the incident.
- Helping to undertake the post-incident review and identify lessons learnt in order to improve the management of future incidents.

With experience, an IC will develop the skills to carry out these actions almost intuitively in a short time. The 'novice' IC may need to use aide memoires, particularly the decision making model, introduced in the mid-1990s, and more recently the decision control process – the DCP in current UK use.

Decision support tools

There are a number of models that can be used to support the process of incident assessment, and these will include the Decision-Making Model (DMM), the Decision Control Process (DCP), and the Joint Decision Making Model (JDM – produced as part of the JESIP Project), all of which are in use in the UK, as well as a variety of methods and systems used in North America and elsewhere. Later chapters will look at some of these models and compare the benefits of each.

Essentially, all processes follow a 'Plan, Do, Review' cycle (sometimes 'Plan, Do, Check, Act'), to which refinements have been incorporated to assist resolving operational incidents. All models broadly incorporate the following steps:

a. Information gathering and initial risk assessment.

b. Deciding priorities and setting strategic objectives.

c. Determining the tactics required to satisfy the strategic objectives.

d. Resourcing the plan.

e. Implementing the tactical response.

f. Reviewing progress against milestones.

While it is not the intention to examine these aspects in detail at this stage (they will form part of subsequent chapters of this book), it is worth looking at some of the key features within these areas to start framing the operational models that will be discussed later.

Managing Incident: decision-making model

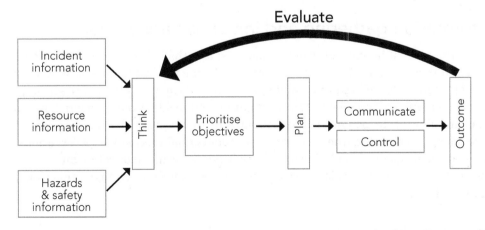

Figure 3.5: Decision making model
(Reproduced from National Operational Guidance (NOG))

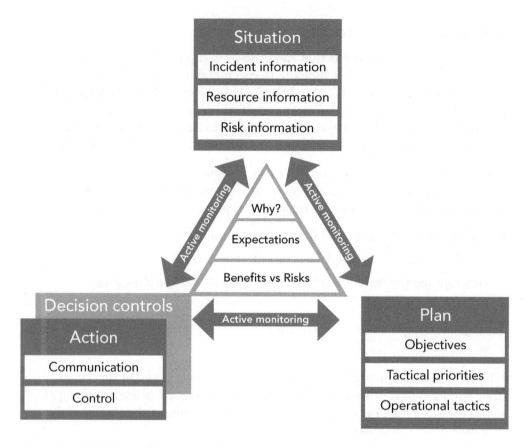

Figure 3.6: The UK Decision Control Process (DCP)
(Reproduced from NOG)

Information gathering and initial risk assessment

Information about an incident can be separated into that which was known before
the event occurred and that which is available during the incident itself. Wherever
possible, knowledge of high-risk buildings within station area, division or service,
should be regularly accessed so that firefighters and incident commanders have an
understanding of the potential risks and hazards that a particular building may
pose. Alongside details gathered during the mobilisation process and observations
made at the incident, this pre-incident knowledge can confirm or refute any
assumptions about the incident through the process known as triangulation (see
Chapter 6 for more details).

Pre-incident knowledge

Knowledge of the type of premises, the processing and activities being undertaken
there, and the associated hazards and risks are all key elements in being able to

manage an incident effectively. Information about premises can be of a generic or a building-specific nature. Generic information is readily available on the risks and hazards associated with common types of structures such as small dwellings, and smaller industrial and commercial premises including shops, factories and offices.

Larger, more complex buildings, or those with unusual features, those undertaking high-risk activities, or buildings storing or manufacturing large quantities of materials are more than likely to have additional information requirements for a responding fire and rescue service. This additional information, sometimes called site specific risk information (SSRI) is likely to include:

- building details including methods of construction
- the location of utilities
- availability and quantity of water supplies
- the presence of fixed installations including automatic fire suppression and ventilation systems
- environmental considerations
- community hazards
- potential loss
- an indication of initial actions and considerations to be taken by the first attending IC
- the relevant tactics to be used when possible.

This information is usually accompanied by a site or building plan indicating key features, including access and rendezvous points.

This information can be stored in hard copy format or electronically on a central server and transmitted to a mobile data terminal (MDT) on attending vehicles. For premises with even higher risks, more detailed plans may be available such as those produced in the UK under the Control of Major Accident Hazards (COMAH) Regulations (2015) for both onsite and offsite activities.

Information about higher risk sites or properties can also be gained by firefighters undertaking intelligence gathering inspections or familiarisation visits. It has been shown that familiarity with a risk greatly enhances operational effectiveness and the safety of crews at actual incidents.

'Live' sources of information

During an incident, information may be available from a number of sources including the presence of an occupier (or responsible person), witnesses to the causation of the event, on-site information boxes for the exclusive use of the fire and rescue services and other emergency services, as well as personal observation by the IC and attending firefighters.

Having considered information about the incident, the risks and hazards and other operational circumstances, the next phase is set strategic objectives objectives and identify tactical priorities.

Box 3.2: Lack of knowledge of layout and construction

A fire occurred in a large chicken processing factory (130m x 140m). It broke out above a fridge within the centre of the building and that set fire to the ceiling lining above. The roof had been renovated within the last six months and was unventilated apart from the ventilation ducts. The suspended ceiling was insulated with rigid polyurethane foam slams sandwiched between two layers of steel. There was a large void above the suspended ceiling and there was no fire resisting subdivision of this void.

Two firefighters with breathing apparatus were tasked with carrying out a reconnaissance of the main compartment and to check the roof space, if possible. Within ten minutes of their entry a message indicating that the team were in distress was received and a 'BA emergency' was declared. Several breathing apparatus teams entered the building and located the missing team members, trapped beneath the collapsed ceiling structure. Due to the intensity of the fire and the threat of exploding cylinder screws, they were forced to withdraw and it was some time later that the bodies of the firefighters were recovered.

The Health and Safety Executive identified a number of areas where the fire brigade was deficient, and served improvement notices on the brigade:

1. *'The information held by the Brigade and available to fire crews and officers on hazards associated with the design and materials of construction of buildings is insufficient to ensure… the health and safety of firefighters…'.*

This applied particularly to the fire behaviour of laminated polystyrene insulated panels which delaminated during the fire, the presence of un-compartmented voids and the early collapse of unprotected steel and aluminium structural components.

2. *'…that the understanding by all firefighting personnel of the procedures for entry into buildings when wearing breathing apparatus is not consistent.'*

Lessons learned

■ All firefighters need to have a greater understanding of the fire behaviour of materials and structures.

- All firefighters need to be familiar with significant risks on their turnout areas (both administrative areas and neighbouring stations).
- Fire and rescue services need to be able to collate and transmit this information when required.

Deciding priorities and setting strategic objectives

At any fire or emergency incident, there will a range of both sequential and simultaneous activities that need to be prioritised and allocated as well as being carried out to achieve a successful resolution. The role of IC is to assess these often competing activities and develop a prioritised list of objectives. The main priority objectives are usually cited as:

- To save life.
- To save property.
- To provide humanitarian services.
- To reduce impact upon the environment.

...all while balancing firefighter risk against potential benefits at all times.

Saving life will always be the first priority in any operational environment and is the most time critical objective: saving property, providing humanitarian services and reducing impact on the environment are, to a certain extent, dependent upon an assessment balancing the needs of the incident and wider considerations, and the time available.

Having determined the priorities, the strategy to achieve a successful outcome can be developed. At its simplest level, a strategy maybe one predicated on offensive or defensive modes of operation. With increased levels of complexity, strategy may be more refined and be comprised of multiple elements. Strategy employed at incidents involving large multi-agency deployments are discussed in more depth in Chapter 16.

Determining the tactics required to satisfy the strategic objectives

The tactics required to meet strategic objectives will vary with the type of incident, its scope and scale, the operational context (location, specific demands of the incident), time of day, weather conditions and the time available in which to resolve the incident. The tactics required will be likely to involve the use of a range of equipment by firefighters with different skill sets and experience, selected to achieve the optimum outcome as quickly and safely as possible. When determining and selecting tactical options, a systematic approach to resolving incidents provides a shorthand way of taking action relatively quickly. Recognition primed decision-making (RPD) is one such way in which experienced ICs can make decisions rapidly based on their attendance at previous incidents. More complex incidents, particularly those involving hazardous materials, complex technical rescues or large buildings/structure fires, will need a more considered, analytical approach, where detailed analysis may be necessary before identifying the most appropriate strategic and tactical options. At the most complex or novel incidents, a creative approach to identifying possible solutions may be required. These decision-making styles are discussed in Chapter 7.

With regard to the tactics and sequencing of tasks at actual incidents, this will depend on the actual circumstances but Appendix A (p277) gives a brief refresher and overview of basic operational tactics and procedures for fires, transportation accidents and hazardous material incidents. It is not intended to give comprehensive guidance as this is covered in more detail in national and service documents guidance and policies, but it will give an overview of some of the elementary processes involved in the major categories of incidents.

Resourcing the plan

Once the tactical solutions to a incident have been determined, resources must be requested to enable implementation of those tactics. Resource requirements for a fire will be determined upon a series of factors that the IC must consider:

a. The size of the fire at that time and a projection of how likely the fire is to spread before it can be contained and extinguished.

b. The number of jets, other firefighting implements and media (foam, carbon dioxide, dry powders), and pumping appliances that will be required to control the incident.

c. The number and type of special appliances that may be required to help control the incident. These include aerial ladder platforms, foam tenders, rescue tenders, technical rescue units and rapid response vehicles.

d. The proximity and vulnerability of exposures surrounding the incident which may require separate and/or additional resources to ensure spread beyond the building of origin is limited.

e. The availability and quantity of water available within a reasonable distance, which would help determine whether a water relay using additional pumping appliances or high-volume pumping units may be required.

f. The contents, processes or storage within the building. This will be a major factor in the potential speed and intensity with which the fire can develop. Knowledge of the contents and the relative flammability together with an understanding of the type construction of the building will help the IC understand how far the fire is likely to spread and develop before control may be possible.

g. The height of the building. This should be taken into consideration when estimating assistance: the taller the building, the greater the number of resources that are likely to be required to achieve effective control of the fire. Firefighters will be required to provide the logistic support to allow operations to be to undertaken on the higher floors by carrying breaking in equipment, hoses, lighting etc. Firefighters will also be required to ensure that the extra safety measures required at these incidents, such as file lift controllers, water management, lobby and bridgehead support and tactical ventilation, are in place, as well as maintaining control of inbuilt fire engineering systems and providing cordon control for an extended perimeter due to potential glass and debris risk.

h. Where salvage of equipment and goods is a high priority (particularly in commercial, industrial premises, and often within domestic properties), additional sectors and resources maybe required. When an incident occurs on an upper floor, the potential for salvage from the lower floors is much greater.

i. Buildings with a potential for rapid internal spread through staircases, access openings and shafts with limited fire protection may need extensive internal breathing apparatus operations, which are resource intensive. Furthermore, where atmospheric conditions, including temperature and humidity, are high, working duration for firefighters will be limited or reduced. Rehabilitation of exhausted firefighters may take up to 12 hours and so additional resources may be required to provide release for breathing apparatus wearers.

j. The time of day and weather conditions may have an impact upon the resources requirements. Additional crews may be required to provide additional safety systems such as lighting during the dark hours. Additional relief crews may be required to take account of shorter working durations caused by extremes of heat and cold.

Once a resource assessment has been carried out, an assistance message should be sent as soon as practicable to mobilising control. It is important to note that traditional methods of requesting assistance using vehicle-based quantities (e.g. 'Make pumps 5'; 'Aerial Ladder Platform required' etc.) are no longer the only method of requested assistance. Asset-based mobilising – requesting resources in terms of number of major firefighting appliances, aerial appliances and other specialist vehicles, followed by requests for a defined number of firefighters – is becoming more common as first response vehicles are not necessarily pumping appliances crewed by four or five firefighters, but may instead be rapid response vehicles (RRVs) crewed by two or three firefighters.

Implementing the tactical response

Having set the strategic objectives, developed tactical plans and identified resource requirements, the next stage is to implement the tactical response to resolve the incident. In doing so the IC will need to organise the incident ground which will incorporate both operational sectors and supporting sectors and take into account any actions that are being undertaken by crews already in attendance. Resources will need to be allocated to each sector and sector commanders will need to be given a briefing setting out their tactical objectives and the limits of their authority and operational parameters. The allocation of radio channels should be determined by the incident commander or the command support team (if in place). A timeline should be organised identifying milestones including the dispatch of messages to fire control, the review of risk assessments and anticipated developments of the fire or other incident types, all of which help the IC to monitor and review progress

Reviewing progress against milestones

In order to assess the effectiveness of operational tactics and actions it is important that progress is measured against expectations on a regular basis. The incident commander should have a considered view of how the incident should be developing with time considering the fire dynamics, the likely impact of actions being undertaken by firefighters and the availability of resources arriving at the incident ground. A regular review will enable the IC to assess the success or otherwise of plans which can lead to modifications of tactics, requests for additional resources or the ability to release resources from the incident. Built into the review process is the need to send regular messages detailing progress to enable strategic managers and fire control to manage strategic resources across the service area.

Summary

You should now understand the following terms, ideas and concepts explored in this chapter:

The differences between 'strategy', 'tactics' and 'tasks' and how they are all used at incidents of varying scale and scope.

The actions required of the first attending incident commander and the decision support tools they may call upon to ensure a safe and effective response, including the Decision Control Process and the older Decision Making Model.

The key tasks of the incident commander, including the gathering of information and initial risk assessment, determining the incident's strategic priorities, objectives and tactics, resourcing considerations and implementing the plan.

Measuring the success or otherwise of activities against the plan and its milestones.

Chapter 4: Incident command systems

This chapter looks at some of the fundamentals of incident command and its purpose. It will look at the requirements for effective incident command systems – the '4Cs': command, control, co-ordination and communications, and it will define what hazards are included within the scope of the 'all hazards approach'. It will explore the underpinning legislation that enables firefighters to carry out their duties at task, tactical and strategic levels, and looks at who is responsible for the health and safety and welfare of staff on the incident ground. The chapter also considers that incident command is underpinned by managerial activity away from the sharp end, which is key to successful operational responses, as is an understanding of the different capabilities and skill sets of responding staff.

Introduction

At its simplest, effective incident command provides a method of delivering the best outcome at an incident by saving lives, property and the environment. It uses systems of work that deliver the greatest benefits for the amount of risk faced by firefighters and others who are involved in the response. An incident command system provides a tool to ensure effective command control and co-ordination for a response to an incident, from a minor collision involving one car to major disasters taking place across a large geographical area involving many agencies and several hundred emergency and recovery responders.

To be of value, an incident command system must be understood by everyone, it must be easily applied, and it must be scalable to the incident being attended. All firefighters and ICs need to be cognisant of the range and extent of operational risks and scenarios they face and they need to be able to apply incident command systems in all circumstances.

Over time – 50 or so years – incident command systems have evolved from service-specific procedures to newer, national systems which ensure a common understanding and interoperability of responses across the country, improving outcomes and enhancing safety for firefighters. It is not the intention of this chapter to focus on any one specific model of incident command, but instead to outline the principles of such systems.

The aims of the UK incident command doctrine are:

■ To support safe systems of work through the use of standard procedures including breathing apparatus, incident command systems and the technical systems associated with incident ground operations.

■ To support effective decision making through the use of decision support tools such as the Decision Control Process, the use of incident support systems such as SSRIs, dynamic organisation charts, decision logging tools.

■ To support effective incident resolution through clear leadership skills and behaviours.

■ To create effective teams making the most of the staff resources available and matching skills and knowledge to the requirements of the role, wherever possible.

■ To ensure staff welfare through the use of monitoring systems such as the introduction of an incident safety sector and observers, the provision of welfare arrangement (such as feeding, rest and recuperation facilities) and facilitating relief of crews and officers to maintain the continued effectiveness of staff.

■ To support environmental protection through effective liaison with other agencies and fire and rescue staff to minimise the impact of the incident and firefighting operations and incidents on the environment.

■ To develop business continuity through effective intervention, mitigation of damage, and the restoration of premises and facilities.

■ To encourage multi-agency working, both before and during incidents.

■ To protect organisational reputation through the effective intervention at operational incidents and demonstrating support for our communities before, during and after traumatic events.

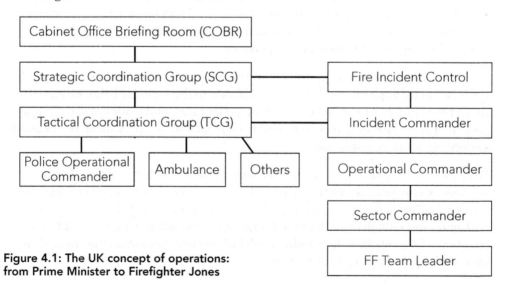

Figure 4.1: The UK concept of operations: from Prime Minister to Firefighter Jones

Incident Command Systems: basics

An incident command system (ICS) is a method of organising incident ground activities to ensure that firefighters work in a co-ordinated and controlled manner that is safe and effective in delivering successful outcomes for the community. An ICS is built around the 'safe person concept', whereby an an organisation selects the right person for the role, gives them appropriate information, personal protective equipment (PPE), the correct tools, training/instruction, and uses safe systems of work and under appropriate supervision. The individual, meanwhile, maintains their competence, works as an effective team member, is self-disciplined and adaptable, vigilant of their own and others' safety and recognises their limitations.

An ICS is composed of a number of key components, including the effective organisation of the incident ground, and using systems of work and tactics appropriate to the nature and extent of current risks. It is commanded by competent officers who have the required training, knowledge and technical skills underpinned by the correct behavioural attributes to effectively command and lead at an incident.

The ICS should be implemented at every incident and will help support the IC in developing an effective incident ground organisation and command structure. Using the ICS framework, the IC will be able to co-ordinate and control the speed and direction of operations using appropriate tactics for the nature of the emergency.

For a single agency only emergency, such as a fire, it is likely that there will be one IC (and a supporting firefighter in the role of command support) who will manage the incident response and who has the final decision-making responsibility and accountability. As incidents grow in size, the command team will expand until it may include an IC, a support team, and sector and functional team commanders. Where an incident grows to sufficient size or complexity, other agencies may be involved and, where necessary, multiple agency command systems will be required using the Joint Emergency Services Interoperability Program (JESIP) protocols.

There can be further augmentation of the incident command system where an incident takes place over wide area or involves multiple sites, such as large area wildfires, or multiple locations of events such as terrorist-related activities. An overview of these types of incidents and their associated command structures incorporated into the UK concept of operations is given in chapter 15.

For the ICS to work effectively, it is important that all command personnel:

■ are adequately trained

■ are familiar with policies and procedures

■ have the necessary competencies for their role

- demonstrate effective command skills
- are confident in their ability
- know who they are responsible for
- know who they need to report to
- know what their operational brief is.

There are a range of components that are required to enable an ICS to function effectively. Sometimes these are known as the four Cs: command and control, co-ordination and communications. These components need to be effectively managed as even the best systems and standard operating procedures and well trained staff will not be able to function effectively if any of these are absent or dysfunctional. All four terms are used widely in discussions about incident command, so we will spend a little time examine these in further detail.

Command and control

The terms 'command' and 'control' originated within military thinking and are often used interchangeably in both spoken and written communication. For the purposes of this book both terms are used together. It has also been said that *'one of the least controversial things that can be said about command and control is that it is poorly understood and subject to widely different interpretation'* (Kenneth Moll), and despite millions of pages having been written on the concept and the act of commanding, no formal definitions exist. In this book, we use the term to refer to *'the exercising of empowered authority to direct resources and personnel to the achievement of common aims and objectives'*.

The IC has a legal authority by virtue of their seniority in their employing organisation and by Sections 7 and 44 of the Fire and Rescue Services Act (2004) to direct firefighters and resources in such a manner as they see fit in order to achieve certain outcomes. While there is no statutory basis for the IC (or fire and rescue service) to have sole responsibility for an incident, at fires, hazardous materials incidents and some other types of specialist rescue events, it is likely that even where other agencies are involved, the FRS will be seen as the lead responding agency. This is likely to mean that the fire IC, is the *de facto* senior person in charge of the incident until the emergency phase concludes and the incident moves into recovery and investigation processes.

Under the Fire Services Act (1947), the most senior FRS officer from the service in whose area the fire occurred was in sole charge for the incident, superseding all other agencies, including the police, until the fire was extinguished and included responsibilities to *'control of all operations for the extinction of the fire, including the fixing of the positions of fire engines and apparatus, the attaching of hoses to any water pipes or the use of any water supply, and the selection of the parts of the premises, object or place where the fire is, or of adjoining premises, objects or places, against which the water is to be directed'.* While this was altered by the Fire and Rescue Services Act (2004), it remains the case that the most senior FRS representative on the incident ground remains the person legally responsible for that incident, associated staff and others, whether or not they have formally taken command.

Box 4.1: A 'confusion' of commanders

A serious fire occurred in a flour mill in 2009 necessitating the mobilisation of 15 pumps, a high-volume pumping unit, a water tower and aerial platforms. A full command team consisting of an area commander and five additional officers formed an incident command structure to manage the incident. A brigade manager self-deployed to the incident to 'Monitor but not take over' the incident and positioned himself in a prominent location on the incident ground. During the incident it was noted that despite taking a 'monitoring' role, the brigade manager was in fact issuing instructions to operational teams and officers without notifying the IC. As a result of these interventions, crews and officers became uncertain of those in charge of the incident and were taking contradictory orders.

Lessons learned

Unless you intend to take over at an incident, do not issue instructions to staff regarding actions they should be taking. If asked for advice, make it clear that you are not the IC, and that direct enquiries should go to the IC.

If you *do* intend to command the incident, notify the current IC of your intentions, take the 'Incident Commander' surcoat from the previous IC and put it on (this may seem a superflous matter but avoids the common problem of having several officers wearing an 'Incident Commander' simultaneously) and, when taking command, make a formal declaration to mobilising control and to those at the incident.

Co-ordination

In any operational context, co-ordination refers to the organisation of the various components of the response or activity in order to enable them to work together effectively. For this to occur number of key requirements must be fulfilled:

- Everybody must be working to the same rules and pulling in the same direction: there can be no 'freelancing' or independent action taken outside the tightly defined parameters set in place by the IC. Freelancing, either the self-deployment of firefighters *of any position in the organisation* to an incident or the undertaking of tasks or activities unauthorised by a member of the command team, distracts resources from the management of the incident, can cause confusion and misdirection on the part of firefighters, and has the potential to play havoc with the incident command structure. This can potentially compromise the command and control of the incident as well as firefighter safety. The principles and key elements of standard operating procedures should be known and understood by all those attending the incident so that no one is put at risk due to a lack of understanding or misunderstanding of procedures

- A chain of command must exist with one person (the IC) as the person responsible for setting direction, strategic objectives and tactical priorities. Having a single point from which direction is given is essential – where multiple instructions, decisions and directions are being created from different people there is the potential for confusion and counterproductive actions on the fire ground which may create unnecessary hazards and risks to firefighters or others: multiple command does not work.

- As incidents grow in size or complexity the supporting organisational structure will increase as the demands for information and additional controls and co-ordination are required. By necessity, the resourcing to support the command, control and co-ordination functions will increase demands for both personnel and equipment. The incident management system is purposely designed to be scalable and to grow with the size and complexity of the incident (see Box 4.2: Levels of command).

Box 4.2: Levels of command

At any incident there will be one of four levels of command within an ICS depending on its scale and severity:

- **Level 1** – Initial command and control operations at a task focused supervisory level (Crew Commander or Watch Commander) or a more senior level at a serious escalating incident. Normally, it is expected that the level 1 commander will manage incidents up to four or five pumps, typically a serious, large housefire, significant road traffic collision, or take command of a sector (both operational or functional) at a larger incident.

- **Level 2** – Intermediate command and control operations at a tactical middle manager level (Station Commander or Group Commander) or a more senior level for large or significant incidents. Depending upon the individual FRS policy, a level 2 commander will be responsible for managing an incident of between 4 and 8 pumps or a sector at a larger incident. In many services, a level 2 commander will be mobilised to smaller incidents to monitor, supervise or act as a monitoring officer to more junior commanders, retaining the option to take command should it become necessary. The type of incident that automatically requires a level 2 attendance may include a 'persons reported' fire, RTC, or Hazmat incident.

- **Level 3** – Advanced tactical command at the largest and most serious incidents, either at the scene or at a remote location. This role is normally filled by a Group Commander or an Area Commander. This role is very often determined by the number of resources required at the incident and will normally be used when an incident has grown to 8-12 pumping appliances. It may be the case that the Level 3 commander may be located at an off-site tactical coordination group as the FRS representative. There is a requirement for tactical co-ordination and of having reached the stage of using a developed command support and a full ICS structure.

- **Level 4** – Strategic command is associated with commanding within a strategic co-ordinating group at a remote location or taking a command role at the most serious incidents in a tactical location at the incident. Normally, this role is undertaken by an Area Commander or a Brigade Commander.

In order to bring the three elements of command, control and co-ordination together it is necessary to have the fourth 'C' in place – communication.

Communication

Without effective communications, there is a risk that the intentions of the IC will not be received or understood at the location of the incident when it is required. Incident communication is not a single entity but is classified into a number of groups:

- The communications between the original caller and mobilising control.

- The communications between the mobilising control and the responding assets, such as vehicles, stations and individual firefighters and commanders, including the initial and subsequent command points/units.

- The command communications network on the fire ground including the tactical channels used to co-ordinate breathing apparatus operations, water supplies, support functions on the incident ground and the face-to-face communications between firefighters, supervisors and other agencies at the tactical end of the operation.

(The nature of communications is dealt with in detail in Chapter 12.)

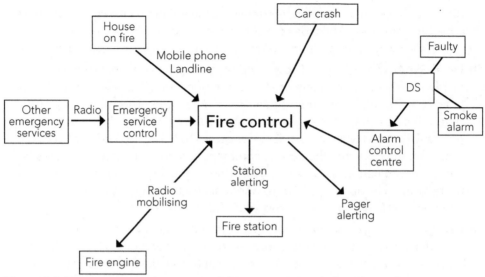

Figure 4.2: The communications network from emergency call to dispatch

The 'all hazards approach'

The concept of the 'all hazards approach' (AHA) originated in the field of disaster and emergency management, but has since migrated to first-line emergency responders. In the wider disaster management context the all hazards approach takes into account an integrated hazard management strategy that incorporates planning for, and consideration of, all potential natural and technological hazards. Within the fire and rescue services it is taken, in essence, to mean that ICs will be trained and have the skills and knowledge to enable them to apply the principles of ICS to deal with most types of incident they may attend.

Needless to say, the more complex the incident and the larger the response required, the seniority of the IC will be correspondingly greater based on their greater level of training and experience in managing larger events. Individual services are likely to have predetermined levels of command attendance based upon the size, complexity and impact of an incident.

Nevertheless, the list of potential incidents the IC can be required to manage is extensive and wide ranging – truly, all hazards are included:

■ Fires.

■ Hazardous materials.

■ Natural hazards including floods, landslips, wildfires.

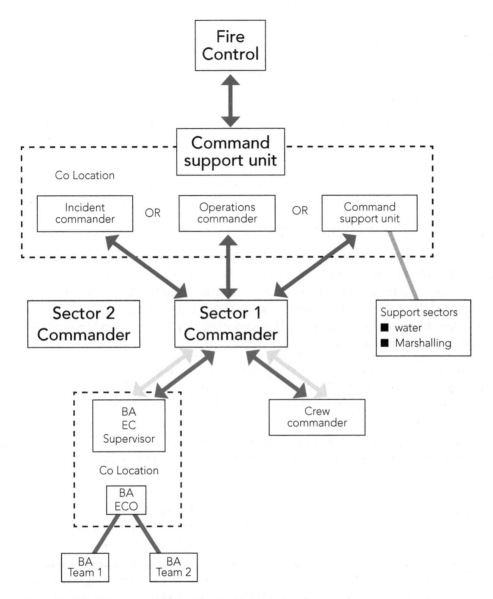

Figure 4.3: Incident ground communications networks

- Road traffic collisions involving entrapment of people or possibly involving hazardous materials and fires.

- Incidents involving aircraft, railways or shipping (both inland, in harbours and out at sea).

- Chemical, biological, radiological, nuclear and explosives (CBRNE) incidents.

- Terrorist events, including armed attacks.

- Rescues from height, below ground and within confined spaces.

- Wide area search and rescue operations for individuals and groups.

- Surface and subsurface water rescue.

- Supporting other agencies in emergencies such as implementing an emergency inoculation programme in the event of an outbreak of communicable diseases, e.g. Ebola.

- A range of activities that may at present be unpredicted and unknown.

Hazards, risks and significance

An understanding of hazards, risks and significance is vital to allow an IC to understand the difference between a minor or trivial risk and one that could have a major impact upon the way an incident is managed. One of the most common problems when talking about risk and hazard is that the two terms are often taken to mean the same thing. It is important for the IC to understand the distinction between the terms.

1. A hazard is something that has the potential to cause harm, damage or adverse health effects upon someone or something. This could be a chemical, an exposed electrical cable, a river, fire or a badly positioned ladder: all have the potential to cause harm.

2. A risk is the likelihood (or chance or probability – all mean the same) that any hazard will cause harm. The likelihood is often expressed as a high, medium or low risk.

The *significance* of a risk can also be a cause of confusion. Almost all human activity has some level of attendant risk. Getting out of bed in the morning can be dangerous – you could trip over the cat, bump into a chest of drawers etc. This could cause harm, and in some circumstances cause serious injury, but the likelihood is so low that it would not be a matter of concern or significance to the individual. Similarly, on the incident ground, there are some hazards that are so common to daily life that they do not merit recording as a matter of significance. Everywhere we walk there is the hazard of slipping, tripping or falling: unless we were walking on a cliff edge or across a field in darkness, then the significance is relatively low. It is important to focus on the serious issues and not be distracted by the trivial or common place hazards and risks.

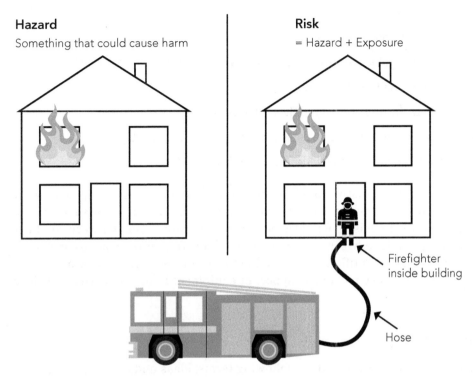

Hazard
Something that could cause harm

Risk
= Hazard + Exposure

Firefighter
inside building

Hose

Figure 4.4: Risk and hazard

Tactical advisers and subject matter experts

The all-hazards approach does not require the IC to have the skills and knowledge necessary to understand all aspects of an incident. The requirement will be to command, control and co-ordinate the response using whatever resources are available to support them. This support may include specialist FRS officers, sometimes called tactical advisers (Tac Ads), and Subject Matter Experts (SMEs), who are able to attend incidents and provide advice on a range of subjects:

■ Hazardous Materials and Environmental Protection

■ Technical Rescue

■ Fire Investigation

■ Fire Safety and Public Protection

■ Wildfire Incidents

■ Command Support

■ Interagency Liaison

- Flood Management

- Urban Search and Rescue

- Chemical, Biological, Radiation, Nuclear and Explosives (CBRNE)

- Firearms Incident Advisers

- National Incident Liaison Officers (NILO)

- Specialist Equipment Advisors including Bulk Foam, CO_2

- Media Liaison and Press Management Teams.

This list is not exhaustive and the number available will depend upon local fire and rescue service needs, the resources available and the area's assessed risks. These specialist advisers will also be able to facilitate liaison between the FRS and any specialist advisers that may be available from other sources, including manufacturers, scientific advisers and premises owners who may be able to furnish information, technical support or equipment to help resolve the incident.

Summary

You should now understand the following terms, ideas and concepts explored in this chapter:

The components of an ICS including command and control, co-ordination and communications ('the 4 Cs').

How the 'all hazards approach' applies to incident command and how ICs can be supported by tactical advisers for a range of incident types.

Chapter 5:
The application of legislation at incident grounds

Introduction

This chapter does not aim to undertake a prolonged study of fire service related legislation, or to ensure that it is fully understood in all its aspects. Rather, it will look at the key issues that are pertinent to the IC at an incident. Obviously, to meet the requirements of these laws, pre-incident actions and activities – such as carrying out Site Specific Risk Information (SSRI) gathering visits and developing standard operational procedures (SOPs) – are required, and these will be discussed. However, this chapter will focus on the law as it impacts upon the decision making processes and thinking on the incident ground. Guidance, including that issued by the HSE, and concepts including operational discretion, are examined in greater detail in later chapters due to their importance and potential for confusion.

Both the Romans and Normans in Britain introduced laws to prevent fires and to provide a means to reduce their impact. These and subsequent laws were continually amended to take account of events in an attempt to improve outcomes in the future. The UK has been fortunate in that relative legislative stability has meant that the law governing firefighters is consistent and slow to change. It is nevertheless important for the IC to know their duties, powers and limitations under the law. It is equally important that they have a working knowledge of other laws that may have an impact on the way they perform their duties. We also live in an age of digital media, where the public image of a service is more important than ever and ignorance of the law is not a valid defence – it can expose the individual and organisation to criticism, which may reduce public confidence in the service.

This chapter is laid out in four parts – the first deals with primary fire and rescue service legislation – the Fire and Rescue Service Act (2004) (FRSA), the Regulatory Reform (Fire Safety) Order (2005) (RR(FS)O) and the Civil Contingencies Act (CCA) (2004). These lay out the specific requirements, powers and duties of the IC during operational incidents. The second part looks at other acts and regulations that relate to the role of the service in the protection of the environment when undertaking operations, from both its own actions and the actions of others. The third part of the chapter looks at the wider legal framework law as it relates to the liability of the FRS when errors are made, and it provides case examples to demonstrate the potential pitfalls which ICs should take all reasonable precautions to avoid. This part also gives an overview of the Corporate Manslaughter and Corporate Homicide Act (2007) and the Human rights Act (1998) as they impact on the FRS.

Given the nature of the role of the FRS, the impact of health and safety legislation on the incident ground and how this may have an effect upon the IC's considerations is significant, and will be considered as part of chapters 10 and 11, which consider all aspects of safety on the incident ground.

FRS specific legislation

The Fire and Rescue Services Act (2004)

The Fire and Rescue Services Act (2004) in England and Wales (the Fire Scotland Act (2005) and the Fire and Rescue Services [Northern Ireland] Order (2006) in those respective countries of the UK) replaced the Fire Service Act (1947) with the intention of recognising the wider role that the service had been playing in community safety, specifically road and domestic fire safety, road traffic accidents and other emergencies, as well as the existing duty to extinguish fires. Part 2 of the Act covers the functions of a fire and rescue authority.

Functions of a fire and rescue authority (Part 2 FRSA)

Section 7 sets out requirements to fulfil the core functions of fire-fighting. They are:

1. *A fire and rescue authority must make provision for the purpose of—*
   a. *extinguishing fires in its area, and*
   b. *protecting life and property in the event of fires in its area.*
2. *In making provision under subsection (1) a fire and rescue authority must in particular—*
   a. *secure the provision of the personnel, services and equipment necessary efficiently to meet all normal requirements;*

b. *secure the provision of training for personnel;*

c. *make arrangements for dealing with calls for help and for summoning personnel;*

d. *make arrangements for obtaining information needed for the purpose mentioned in subsection (1);*

e. *make arrangements for ensuring that reasonable steps are taken to prevent or limit damage to property resulting from action taken for the purpose mentioned in subsection (1).*

This clearly sets out the duties of a FRA to meet the needs of the community in respect of outbreaks of fire within a geographic location. Section 7(2)(d) is of particular relevance as far as incident ground operations are concerned: the provision of and access to information about risks within a community or building or area. This will support the IC in planning because it should mean that knowledge of building type, construction, occupancy, specific problems and processes undertaken within the building, firefighting facilities, water supplies and specialist information is available. ICs should have a knowledge of premises that have such risk information and how it is to be accessed at incidents. This may also assist the IC in meeting the requirements of Section 7(2)(e) when assessing the potential for spread of the fire and the location of high value materials in the event that a salvage strategy is used to protect property.

Other types of information that may be available for use by firefighters responding to emergencies include:

- Site specific risk information (SSRI) for higher risk locations.

- Site specific salvage plans at heritage sites and other high value locations comprising all or specific part of a building.

- Control of Major Accident Hazards (COMAH) Regulation (2015): on site and off site plans are pre-prepared plans for high risk industrial sites. They are required to be provided under legislation which is the responsibility of the premises occupier but can form a useful tool for planning operational responses to an incident both before and during an incident.

- A responsible person's fire action plan produced for fire safety and damage limitation purposes under the RR(FS)O may help indicate the premises' evacuation scheme and 'stop lines' (i.e. fire resisting partitions and doors, water drenchers, etc.) which can be used to support containment of fires.

Sections 8 and 9 of the act deal with road traffic accidents and emergencies. It sets out requirements for services to make provision for rescuing people and protecting people from serious harm in the context of road traffic accidents (RTAs) and emergencies other than fires and RTAs. The Home Secretary will determine which of these 'emergencies' are appropriate for FRA response. Section 10 gives the Home Secretary the ability to direct FRAs to take (or not take) actions in relation to specific incidents, and Section 11 gives an FRS the power to take any appropriate action where a situation occurs that causes or is likely to cause one or more individuals to die, be injured or become ill, or to cause harm to the environment.

By virtue of these duties, an IC may find themselves facing a whole range of situations that they may have reasonably expected to be prepared for, but also events for which they have not prepared but for which they will be expected to take mitigating actions to reduce community harm. This re-emphasises the need for an 'all hazards approach' to be taken when command training. Reinforcement schemes (s13 of the FRSA, mutual aid arrangements) and arrangements to obtain support from other organisations that employ firefighters (s15 of the FRSA, which can include works fireteams and airport rescue and firefighting services (ARFFS)) help to provide a robust national response for larger incidents. However, from a practical point of view this does create the requirement for the 'host' FRS (the body that is hosting the supporting resources) to understand policies, procedures and protocols and, especially, the capabilities and limitations of those organisations providing the support.

Powers of firefighters in an emergency

Under Section 44 of the FRSA, an employee of the FRA may do anything they 'reasonably believe to be necessary' if a fire (or an impending fire), road traffic accident or other emergency has occurred (or is about to occur) and discharging of a function of the FRS is possible. It is also permissible to take such action to prevent or limit damage resulting from the circumstances above. Firefighters may:

- enter a place or premises, by force if necessary, without the consent of the owner or occupier

- move or break into a vehicle without the consent of the owner

- close highways

- stop and regulate traffic

- restrict the access of persons to premises or a place.

Other powers of a firefighter

Under Sections 8(2)(d) and 9(2)(d) of the FRSA, a firefighter can at any reasonable time enter premises to gather risk information to enable the service to carry out its function with respect to road traffic accidents or other emergencies.

Section 45 of the FRSA permits an 'authorised officer' to enter any premises other than dwellings at any reasonable time to obtain information to assist in firefighting, road traffic accidents or other emergencies or investigate the cause and development of fires. Where a fire has occurred in a private dwelling (whether occupied or not), a firefighter may not enter unless 24 hours' notice in writing has been given unless invited by the occupier to do so. Similarly, entry is restricted to buildings where there has been a fire if the premises are occupied or occupied as a private dwelling immediately before the fire unless 24 hours' notice in writing is given. Specialist fire investigation officers will be familiar with those restrictions but the IC should bear in mind that there is no carte blanche to enter a building at any and all times.

Section 46 specifies a number of supplementary powers that exist which allow a firefighter to take with him a person or equipment that may be necessary to support investigations, or require any person on the premises to provide information about the facilities, documents, test records or other items that may be reasonably requested. It may also be necessary to carry out tests on substances or materials found on site. All evidence seized should be managed in accordance with rules on the seizure and chain of evidence rules currently applicable. It is important to remember that where a premises is unoccupied or temporarily unusable, the premises must be secured against unauthorised entry (in the same condition in which they were found).

Emergency Workers (Obstruction) Act (2006)

Under the Emergency Workers (Obstruction) Act (2006) it is an offence for a person to obstruct or hinder a firefighter responding to emergency circumstances without a reasonable excuse (an actual or suspected emergency). These emergency circumstances include those that are causing or likely to cause serious illness or injury, serious harm to the environment, serious harm to any building or other property, or worsening of any injury illness or harm, or which are likely to cause death. 'Obstruction' includes by physical means or actions other than physical, or where action is taken against equipment, vehicles or items that are intended for use in dealing with the emergency. Conviction of this offence will lead to a level 5 fine on the standard scale (£5,000 in 2018).

Where access to property is prevented by someone, then local instructions should be to request the attendance of police, stating the reasons for the request and the relative urgency with which they are required to attend. The prosecution will be managed by the police service working with the Crown Prosecution Service.

Civil Contingencies Act (2004)

Following a series of national emergencies during the first five years of the 21st-century (namely, fuel protests in 2000, the foot and mouth crisis of 2001, flooding between 1990 and 2004, and firefighters'national industrial action between 2002 and 2004), various strands of emergency and disaster legislation were consolidated into the Civil Contingencies Act (2004) (CCA).

Under the CCA, an emergency is now defined as:

- an event or situation which threatens serious damage to human welfare in a place in the UK
- an event or situation which threatens serious damage to the environment of a place in the UK
- war, or terrorism, which threatens serious damage to the security of the UK.

Additionally, to constitute an emergency, an incident or situation must also pose a considerable test for an organisation's ability to perform its functions. The common themes of emergencies are: the scale of the impact and possibly the duration of the event or situation; the demands it is likely to make on local responders; and the exceptional deployment of resources.

Civil contingencies events can be of long or short duration, take a significant period of time to develop ('rising tide' events such as a pandemic or drought) or are 'sudden impact' or unplanned events (such as a major industrial accident or transportation incident involving large numbers of casualties). Responding organisations are categorised by the CCA into category 1 responders and category 2 responders, each with different levels of responsibilities under the act.

Category 1 responders
A category 1 responder is defined as a body that is most likely to be at the core of the response to most emergencies. Category 1 responders are subject to the full range of civil protection duties in the act. Category 1 responders include:

- Local authorities
- Police forces
- Fire services
- Ambulance services
- Her Majesty's Coastguard
- National Health Service organisations
- Public Health England

- Port health authorities
- The Environment Agency, Scottish Environment Protection Agency, Northern Ireland Environment Agency

Category 1 responders are required to:

- assess the risk of emergencies occurring and use this to inform contingency planning
- put in place emergency plans
- put in place business continuity management arrangements
- put in place arrangements to make information available to the public about civil protection matters and maintain arrangements to warn, inform and advise the public in the event of an emergency
- share information with other local responders to enhance co-ordination
- co-operate with other local responders to enhance co-ordination and efficiency
- provide advice and assistance to businesses and voluntary organisations about business continuity management (local authorities only).

Category 2 responders

Category 2 responders include:

- Electricity distributors and transmitters
- Gas distributors
- Water and sewerage undertakers
- Telephone service providers
- Network Rail
- Train operating companies and freight operating companies
- Transport for London
- London Underground Limited
- Highways agency or strategic highways company
- Airport operators
- Harbour authorities
- NHS strategic health authority
- Health and Safety Executive
- Voluntary agencies.

Category 2 organisations are 'co-operating bodies'. They are less likely to be involved in the heart of planning work, but will be heavily involved in incidents that affect their own sector. Category 2 responders have a lesser set of duties – co-operating and sharing relevant information with other Category 1 and 2 responders.

Local resilience forums

Category 1 and 2 responders come together to form 'local resilience forums' (generally based on police service areas), which aim to help co-ordination and co-operation between responders at the local level.

The wider impact of major incidents are addressed in Chapter 16.

Fire and rescue service responsibilities

As a category 1 responder, the fire and rescue service is likely to be involved in most sudden impact emergencies – explosions, transport accidents, major fires etc. – where the response phase of the incidents is relatively short, perhaps only several hours. The process recovery is generally left in the hands of the 'non-emergency' responders such as local authorities and other category 2 responders whose objectives will be to restore the community and environment to something like its pre-disaster conditions. The fire and rescue services responding to an emergency will generally have short-term objectives, including:

- saving and protecting human life
- relieving suffering
- Containing an emergency by limiting its escalation or spread and mitigating its impacts
- safeguarding the environment
- protecting property, where practical
- supporting other agencies.

For 'rising tide' events, the fire and rescue service may not necessarily have a leading response role but may be able to provide resources to support other agencies. Such rising tide incidents include things like spring tide flooding or fluvial flooding, which can be predicted and which follows known and predictable cycles. This allows agencies, including the FRS, to prepare before the event and undertake preventative action including evacuations of the population, the preparation of barriers and clearing flood relief ditches and culverts.

The fire and rescue services have a wide-ranging role at emergencies and wider disasters. These roles and functions are based upon existing capabilities and follow one of the basic doctrines of UK emergency planning: responders should use existing skills, capabilities and assets to support the response to the emergency rather than developing new skills for that emergency. Therefore, the primary activities of a firefighters in an emergency are, among other things:

- to extinguish any fire and rescue anyone trapped by fire, wreckage or debris
- prevent further escalation of an incident by controlling or extinguishing fires, rescuing people, and undertaking other protective measures
- deal with released chemicals or other contaminants in order to render the incident site safe, or recommend exclusion zones
- assist other agencies in the removal of large quantities of flood water
- assist ambulance services with casualty-handling, and the police with the recovery of bodies.

In some areas, there are agreements between fire and rescue and the police for controlling entry to cordons. Where this is the case, fire and rescue are trained and equipped to manage gateways into the inner cordon and will liaise with the police to establish who should be granted access and keep a record of people entering and exiting.

Where required, a fire and rescue authority will undertake mass decontamination of the general public in circumstances where large numbers of people have been exposed to chemical, biological, radiological or nuclear substances. This is done on behalf of the NHS, in consultation with ambulance services.

While powers existing under the Fire and Rescue Services Act (2004) can be applied at a major emergency, the fact that such emergencies invariably involve a wide range of agencies, means that the necessary duties and powers of responding agencies will be the subject of more detailed discussions as part of the joint decision-making protocols agreed as part of the 'Joint Emergency Services Interoperability Program' (JESIP – to be discussed in more detail in chapter 15).

Regulatory Reform (Fire Safety) Order (2005): incident ground implications

The Regulatory Reform (Fire Safety) Order (2005) (RR(FSO)) has little direct impact on operational activity but where inspection activities are carried out by firefighters (which are specifically governed by the RR(FSO)) it will assist in directly gaining knowledge of the layout and compartmentation, location of firefighting media, fire engineering systems and facilities built into a building, sprinkler installations and other facilities for use during a fire. Such pre-incident knowledge can be invaluable during incidents and help support a successful resolution.

Environmental law and firefighting

Protecting the environment is a key aspect of the firefighter's role, but during operational activities it is possible that there will be potential for harm to be caused to the environment. It is therefore important for the IC to have a broad understanding of what is permitted and what are the limitations of operational practices with regard to environmental law.

The law in respect of the environment is constantly changing, but there are some fundamentals that the IC needs to bear in mind as part of the strategic and tactical planning process. During the information gathering stage of an incident, which supports operational incident management, any environmental risks need to be assessed including the storage of hazardous materials and the use of operational media such as foam making equipment and the volume of water likely to be required to control an incident. These should be noted and assessed as soon as possible during an incident using the National Environmental Risk Assessment performa, developed by the National Operational Guidance Programme and the Environment Agency (see Appendix C on page 287).

The four aspects to consider with respect to the environment are:

- water quality of both surface and ground waters, and also coastal waters
- sewerage systems
- land and soil
- waste, and in particular hazardous waste.

Environmental Permitting (England and Wales) Regulations (2010)

Surface water and groundwater protection

Under Regulation 38(1) of the Environmental Permitting (England and Wales) Regulations (EPR), it is an offence to cause or knowingly permit water discharge activity or groundwater activity unless provided with a valid permit or exemption. Water discharge activities include discharging poisonous, noxious or polluting matter or solid waste into inland freshwater, coastal waters or territorial waters; discharging trade or sewage effluent into inland waters as detailed above; cutting or uprooting vegetation in any inland freshwater without taking reasonable steps to remove it from the water course.

'Causing' includes undertaking an active operation or failing to take action (such as maintaining a piece of equipment). Where an IC allows fire effluent to discharge into a watercourse without taking any mitigating action or notifying the relevant environment agency, they are guilty of an offence.

With respect to groundwaters, activity includes the discharge of a pollutant (e.g. firefighting foam or fire effluent) that results in, or might lead to, a direct or indirect input to groundwater. This includes a wide range of discharges, such as cement powder, chemicals or fertiliser in solid or liquid form.

Regulation 40 provides defences for FRSs where their actions have resulted in pollution. It is essential for ICs to know these permissible defences, which are:

- the discharge was made in an emergency to avoid danger to human health
- the person took all steps as were reasonably practical for minimising pollution
- particulars of the actions leading to pollution were furnished to the regulator as soon as reasonably practicable after the pollution occurred.

Essentially, as soon as a potential problem is recognised, the IC should arrange for actions to be taken which prevents or reduces further pollution, mitigates any existing pollution where possible and notifies the relevant environmental protection agency as soon as possible.

The Environmental Damage (Prevention and Remediation) (England) Regulations (2015) (EDR)

The Environmental Damage (Prevention and Remediation) Regulations (2015) (EDR) addresses environmental damage and adverse effects on Sites of Special Scientific Interest (SSSI) or on the conservation status of species' habitats protected by other EU legislation, adverse effects on surface or groundwater contamination of land by substances, preparations, organisms, micro-organisms that resulted in a significant risk of adverse effects on human health. The fire and rescue service, in order to comply with the EDR, must take practical steps to prevent environmental damage as a result of its activities, both where there is an imminent threat of damage occurring and where there is a threat of further damage. It should notify the relevant enforcing authorities as soon as possible to ensure that the optimum solutions for mitigation are identified and implemented. Where damage has been caused or worsened as a result of FRS activity, then it may be guilty of committing an offence. If the event was authorised in writing, the damage was the result of an act of a third-party (i.e. the person causing the pollution), or the damage was caused as a result of compliance with instructions given by a public authority (e.g. the Environment Agency), then any liability for remedial costs may be defended against.

The wider legal framework and impact upon the FRS

FRS liability for negligence

It is important that ICs are aware of their obligations under common law to ensure that the actions they take or instigate do not result in harm or loss for which they could be held accountable, either personally or as representatives of their FRS (or their ultimate employers – the fire and rescue authority). There are several basic concepts with which the IC should therefore be familiar. These are:

- the responsibilities and duties of the FRS
- the notion of vicarious liability
- negligence
- the duty of care.

While legal cases related to the fire and rescue service are limited in number, they set precedents that all firefighters working in the FRS should be aware of – both to understand the limits of their powers, as this can impact on what can be achieved by firefighters, the operational competence they need to demonstrate, and the understanding of what is permissible. In so doing, firefighters will be enabled to take actions, the avoidance of which may lead to an accusation of risk aversion or risk avoidance.

Responsibilities and duties of a fire and rescue authority

The duties and powers detailed in the FRSA 2004 (see above) can be summarised as making provision for the purpose of extinguishing fires and protecting life and property; entering premises and doing anything reasonably necessary for the purpose of extinguishing fire or protecting property. This includes closing highways and stopping traffic and breaking into premises if it is believed a fire has broken out, and taking any action necessary to extinguish the fire – with or without the consent of the owner or occupier. These are wide ranging and extensive powers which put firefighters in a position of trust – a position that should not be misused or abused.

Vicarious liability

Essentially, the concept of vicarious liability applies to all employees who, in the course of their employment (i.e. carrying out duties under the FRSA 2004), act negligently in such a manner that damage is caused to or losses are incurred by a third party. A 1952 ruling in *Kilboy v South Eastern Fire Area Joint Committee* (in which a firefighter threw a line which struck the claimant in the eye) held that *'a fire officer is vicariously liable for acts of negligence committed by its members of its fire brigade acting in the course of and for the purposes of, their duties'*. The body that will therefore be sued in the event of losses will be the FRA (the employer) and not the individual employee. Where an action by an employee is deliberate and not taken as part of his/her duty, then it may be the employee who is held personally liable for the act and not the employer.

Claims in negligence: the basics

To establish negligence, any claimant (an aggrieved party) needs to establish three things:

1. That the defendant (the FRA) owed him/her a duty of care (which includes avoidance of the damage that was caused).

2. The defendant was in breach of that duty.

3. The claimant has suffered damage as a result of that breach.

Duty of care

Whether the defendant owes a duty of care is conditional upon the establishment of three elements:

1. The **foreseeability** of damage (the real and substantial risk or chance that something like the event might occur).

2. **A relationship of 'proximity' or 'neighbourhood'** between the parties such as that which exists as soon as a firefighter takes command of an incident or is responsible for an activity in part of the incident ground.

3. That it is **fair, just and reasonable** that the law should impose a duty in the circumstances giving rise to the case.

Case law

A number of cases have helped establish the current understanding of the liability of FRAs over the last 25 or so years, which serve as guidance for ICs. These cases are:

- *Capital and Counties PL v Hampshire County Council.*

- *Digital Equipment Co Ltd v Hampshire County Council.*

- *John Monroe (Acrylics) Ltd v London Fire and Civil Defence Authority.*

- *Church of Jesus Christ of Latter Day Saints (Great Britain) v West Yorkshire Fire and Civil Defence Authority.*

- *Daly v Surrey County Council.*

Capital and Counties PLC v Hampshire County Council (1997)

In March 1990 a fire broke out in a premises in Basingstoke. At 10.23am an automatic sprinkler system began to operate in the roof space at about the same time the FRS arrived. Due to the impact that the sprinklers were having on firefighting operations, the IC instructed that the sprinklers be shut off. At that stage of the incident, firefighters had not yet located the fire. At 10.55 they located it but within 15 minutes, no longer controlled by the sprinklers, it had involved the whole of the roof space and caused the roof to collapse. By 12.10pm the building was a total loss.

At the Court of Appeal it was determined that, *'Where the rescue / protective service itself by negligence creates the danger which caused the plaintiff's injury there is no doubt in our judgement that that the plaintiff can recover'*. (Note: 'injury' in this instance means the loss of the building, and not a personal injury). The reason given by the fire officer for turning off the sprinklers was that, in addition to the problems they caused for firefighting operations (impeding firefighters' vision and progress in the building), it was believed that turning off the sprinklers would reduce damage to the computers. The court of appeal considered the officer to be negligent as the decision he made was one that *'no reasonably well-informed and competent fireman could have made'*. It is a matter of fact that even trainee firefighters are taught not to turn off an operating sprinkler system without instructions and only then when the fire is out and covered by other extinguishing media. The case cost the council over £16 million.

John Monroe (Acrylics) Ltd v London Fire and Civil Defence Authority (1997)

During filming for an episode of *London's Burning* (a TV drama about fictional London firefighters) in the early 1990s, special effects caused a deliberate explosion on waste land that sent burning debris into the claimant's yard. Smoke was seen and the fire brigade arrived to find a fire on the waste land extinguished, and after 20 minutes they left. A fire then broke out in the claimant's premises some hours later and it was alleged that the fire brigade had been negligent in not inspecting the premises to ensure all risks had been eliminated. The court of appeal rejected the appeal and ruled in favour of the FRS, judging that, '*the fire brigade was not under a common law duty to answer the call for help, and are not under a duty to take care to do so. If, therefore, they fail to turn up, or failed to turn up in time, because they have carelessly misunderstood the message, got lost on the way or run into a tree, they are not liable*'.

Jesus Christ of Latter-Day Saints v West Yorkshire FCDA (1997)

In October 1992, West Yorkshire FRS sent 15 pumps to deal with a serious fire in a church building. While attempting to secure water supplies they identified a potential seven hydrants around the premises. Four failed to work and the other three were not found, or not found in time to be of use. Water was eventually sourced from a dam 800m away. The claimants held that the FRS was negligent and in breach of a statutory duty – if they had fought the fire more efficiently, less damage would have been suffered. The claim of negligence was rejected because, apart from the failure to find hydrants near the chapel, none of the allegations related to fighting the fire and the failure was not to place equipment in a location to enable a fire to be fought. The statutory breach was also rejected as the duty regarding hydrants was not specific and limited – rather, it is, '*more in the nature of a general administrative function of procurement*'.

Daly v Surrey County Council (1997)

In 1991, a worker at a construction site in Guildford, Surrey, was digging out a trench when it collapsed, trapping him. The fire brigade was called and arrived a few minutes later. By this time, the worker's colleagues had started to use mechanical diggers in an attempt to rescue him. The IC instructed them to stop digging to prevent a further collapse, which might endanger themselves and their trapped colleague. Upon being rescued, the casualty died. It was claimed that the fire brigade had been negligent in stopping the workers' rescue attempt. The court rejected this claim but stated that the same considerations as in *Capital and Counties PLC v Hampshire County Council* (above) should be applied to circumstances where firefighting is not involved. The instruction was a 'positive act' which could, in principle, give rise to a claim against the service.

In the litigation environment of today, civil authorities are sometimes seen as 'cash-rich targets' for claimants using legitimate or (sometimes) spurious arguments to gain rewards. If nothing else, ICs need to be aware of the limitations of their authority and the potential for claims of negligence to be issued against their FRS. It is therefore essential that decisions made by ICs, especially key decisions which may have a significant impact on life or property safety, are recorded and that they provide a detailed rationale and justification for making those decisions.

Corporate Manslaughter and Corporate Homicide Act (2007)

Before 6 April 2008, it was possible for a corporate entity, such as a company or FRS, to be prosecuted for a wide range of criminal offences, including the common law offence of gross negligence manslaughter. However, in order for the organisation to be guilty of the offence, it was also necessary for a senior individual who could be said to embody the company (also known as a 'controlling mind') to be guilty of the offence. This was known as the identification principle and was difficult to apply in many instances. In addition, where the failings lay with a number of individuals, prosecutors could not aggregate the failings of everyone involved.

On the 6 April 2008, the Corporate Manslaughter and Corporate Homicide Act (2007) (CMCHA) came into force throughout the UK. In England, Wales and Northern Ireland, the new offence created by this act is called 'corporate manslaughter', and in Scotland it is called 'corporate homicide'. CMCHA overcame the problems of accountability and aggregation by providing a means of accountability for very serious management failings across the organisation, and there is now a liability for organisations which could never previously be prosecuted for manslaughter.

S1 (1) of the CMCHA states that, *'an organisation to which this section applies is guilty of an offence if the way in which its activities are managed or organised –*

- *causes a person's death; and*
- *amounts to a gross breach of a relevant duty of care owed by the organisation to the deceased.'*

An organisation is guilty of an offence only if the way in which its activities are managed or organised by its *senior management* is a substantial element in the breach.

This offence is indictable only[1], and on conviction the judge may impose an unlimited fine.

The elements of the offence to be proven are:

■ The defendant is a qualifying *organisation*.

■ The organisation *causes* a person's death.

■ There was a *relevant duty* of care owed by the organisation to the deceased.

■ There was a *gross breach* of that duty.

■ A substantial element of that breach was in the way those activities were managed or organised by *senior management*.

■ The defendant must not fall within one of the *exemptions* for prosecution under the CMCHA.

Partial exemptions

Section 6 of CMCHA clarifies the situation in respect of the emergency services. The offence does not apply to the emergency services when responding to emergencies. This does not exclude the responsibilities these authorities are to provide a safe system of work for their employees or to secure the safety of their premises. Emergency circumstances are defined as those that are life-threatening or which are causing, or threaten to cause, serious injury or illness or serious harm to the environment or buildings or other property. This partial exemption applies to:

1. A fire and rescue authority in England and Wales.

2. The Scottish Fire and Rescue Service.

3. The Northern Ireland Fire and Rescue Service Board.

4. Any other organisation providing a service or responding to emergency circumstances either:

   a. in pursuance of arrangements made with an organisation within paragraph (a), (b) or (c)

   b. (if not in pursuance of such arrangements) otherwise than on a commercial basis.

---

1   This is the most serious category of offence and is dealt with by a Crown Court. Common indictable only offences are murder, manslaughter, causing really serious harm (injury) and robbery.

According to the explanatory notes to the Act:

'The effect of exemption is therefore to exclude from the offence matters such as the timeliness of a response to an emergency, the level of response and the effectiveness of the way in which the emergency is tackled. Generally, public bodies such as fire authorities and the Coastguard do not owe duties of care in this respect and therefore would not be covered by the offence in any event. In some circumstances this may however be open to question. The new offence therefore provides a consistent approach to the application of the offence to emergency services, covering organisations in respect of their responsibilities to provide safe working conditions for employees and in respect of their premises, but excluding wider issues about the adequacy of their response to emergencies.'

The Human Rights Act (1998)

In the UK, human rights are protected by the Human Rights Act (1998) (HRA). Public authorities must follow the act. A schedule of articles was published which defines the rights and freedoms enshrined in the act. If a public authority such as the fire and rescue service has breached anyone's human rights, they may be able to take action under the act. While it is not expected that the HRA is at the forefront of a firefighter's thoughts at most operational incidents, it is worth considering that when engaging with the public, some articles may be relevant to a situation, particularly when life may be at risk and decisions need to be made about a particular course of action. S6 (1) states that, subject to specific exceptions,
'It is unlawful for a public authority to act in a way which is incompatible with a Convention right.'

The use of the JESIP Joint Decision Making Model (JDM) requires consideration be given to duties and powers of the FRS (and other agencies) and it is that this point the HRA articles should be considered to ensure that the requirements are not breached.

The most relevant articles for the FRS are:

Article 2: Everyone's right to life shall be protected by law. Where a FRS/FRA as failed to protect a life through the use of prevention or mitigating measures, then the 'right to life' has been violated. Where more than one agency has been involved in the decision or lack of decision that led to a death, then all participants in the process may be liable.

Article 6: The right to a fair trial may become relevant when dealing with access to a property, obstruction of emergency workers legislation, or during a fire investigation. In both circumstances, any arrest and prosecution will be a matter for the police and crown prosecution service but an awareness of the rights of the individual is of great value.

Article 8: The right to respect for private and family life includes the expectation that everyone has the right to respect for his private and family life, his home and his correspondence and that there shall be no interference by a public authority with the exercise of this right except such as is in accordance with the law and is necessary in a democratic society in the interests of national security, public safety or the economic well-being of the country, for the prevention of disorder or crime, for the protection of health or morals, or for the protection of the rights and freedoms of others. This may become of relevance when seeking to evacuate persons from a premises which has a high risk of a fire or other accident. An order for tenants to evacuate a high-rise dwelling (whether a fire has broken out or not, such as the circumstances where inappropriate cladding has been installed), may well be a violation of human rights under article 8.

Human rights legislation is developing all the time and changes to the status of the UK within Europe may have an impact upon its application. It is likely that the HRA will continue in some form and an awareness of its implications for the FRS, and indeed all of the legislation mentioned above, is an essential for any incident commander.

Summary

You should now understand the following legislation and cases discussed in this chapter:

The Fire and Rescue Services Act (2004) including the functions of a fire and rescue authority, Powers of firefighters in an emergency and other powers of a firefighter.

How the Emergency Workers (Obstruction) Act (2006) may be used to support the resolution of incidents.

The implications and requirements imposed on the FRS through the Civil Contingencies Act (2004) including participation in local resilience forums and its responsibilities in major incidents and emergencies as part of a multi-agency response.

How the Regulatory Reform (Fire Safety) Order (2005) can be used to support operational intelligence gathering.

The potential impact of environmental law on firefighting in particluar regard to the Environmental Permitting (England and Wales) Regulations (2010) (EPR).

Surface water and groundwater protection and The Environmental Damage (Prevention and Remediation) (England)Regulations (2015) (EDR).

The wider legal framework and its impact upon the FRS including the FRS liability for negligence, the concepts of duty of care and of vicarious liability, claims against the service for negligence.

Case law outcomes and their implications for the FRS and incident commanders.

The implications and exemptions of the Corporate Manslaughter and Corporate Homicide Act (2007) and the Human Rights Act (1998) on the FRS, especially with respect to major incidents and events.

Chapter 6: Information gathering and situational awareness

Introduction

Before operational decisions can be made on the incident ground, ICs need to understand the environment in which they are working, the risks they are facing and the tactical options available to them, and they need to develop alternative scenario projections as to how the incident may develop. A comprehensive understanding of what is happening in the immediate and wider environment of the incident enables the IC to formulate a tactical plan to resolve the incident. This chapter will consider the concept of 'situational awareness' and how it applies to the fire and rescue operational environment. It will also consider the key component of situational awareness, which is the gathering of information – its sources, utility and accuracy – which forms the foundation upon which operational decision making is built.

Situational awareness

In its simplest form, situational awareness is about having an accurate perception of a situation as it actually exists. For the IC, this means understanding what is happening on the incident ground by taking information from a number of sources, including personal perception and observation, and triangulating it to build up a complete and accurate picture of what is happening.

The background to situational awareness

The concept of what is now termed 'situational awareness' has been used for millennia within military contexts. Understanding the intentions and disposition of an enemy would (and still does) give military commanders an edge in battle because it allowed them to pre-empt attacks and plan counter-attacks at weak spots. The term itself is believed to have originated during the First World War and was commonly used by United States Air Force pilots in both the Korean and Vietnam wars.

The foundation for the current use of situational awareness was developed by Mica Endsley in the mid-1990s, and her model of situational awareness is among the most widely used across a range of industries – including emergency services, law enforcement and military – to improve operational effectiveness and staff safety. For the FRS, situational awareness is an individual's perception and understanding of the events occurring at an incident and the projection of what may occur in the future. A good understanding of the operational environment will enable the IC to assess the likely impact of various tactical options on the incident and project the effect those actions will have.

Endsley's model of situation awareness has three levels:

Perception: (Level 1) The first stage in achieving situational awareness is to perceive what is happening to the components or elements within the environment. This involves recognising what is happening on the incident ground, identifying cues that may indicate developments in the dynamic of a fire, for example, and monitoring activities that are being undertaken at that incident – the way a building is behaving while on fire, the deployment of resources, the behaviour of occupiers, passers-by and firefighters. This will lead to an awareness of what is happening at that incident in terms of situation elements (objects, events, people, systems, environmental factors) and their current states.

Comprehension: (Level 2) Understanding what is happening and why it is happening is the next step in gaining situational awareness: having seen the visual and audible clues, the mind will start to develop a picture of what is happening and, importantly, *why* it is happening. In this way a narrative of the event will begin to emerge and the IC, having seen the 'what', starts to gain the understanding of the 'why' and 'how'.

Anticipation: (Level 3) Having understood what is happening and why it is happening, it then becomes possible to start projecting forward and anticipating how the incident may develop. Ideally, the IC will identify a number of potential scenarios and consider the requirements to enable that scenario to be controlled and finally resolved. This process is sometimes known as optioneering: the in-depth consideration of various options by assessing how a resolution – resources, equipment, tactics and associated risks – will be obtained and at what cost. Having anticipated a likely or possible development, planning for the control of that eventuality can become crystalised and tactics and tasks allocated in advance. From a practical perspective, the timescales for the projection of an incident's development will depend on the scale and dynamic of the event. An IC may realistically anticipate to control a straightforward bedroom fire within 30 minutes, while a factory fire may take several hours to subdue and larger incidents, such as wildfires or building collapses, may take days to resolve.

Effective situational awareness ensures that the actual situation is reflected in the IC's interpretation of events, based upon information gathered. This allows the IC to plan operational activities effectively, identify how the incident is likely to develop, predict the impact of intervention activities, and help ensure that the correct actions are taken to resolve the incident.

Shared situational awareness

The IC should always aim to have a thorough understanding of the situation facing them, but unless that understanding is shared with the teams operating at the incident then there is a danger that misunderstandings could occur that may lead to unforeseen consequences on the incident ground (see Box 6.1). It is also important that the awareness of the situation is shared with all parties at the incident. This includes 'blue light' agencies, building owners and occupiers, health agencies or any other parties who may be attending the incident. Resolving many larger incidents will be a multi-agency or team effort. The best resolution will occur when all team members have a common understanding and interpretation of events – a shared situational awareness. If team members have different impressions and perceptions of an incident there is the potential for conflicting views regarding the best solutions and other misunderstandings which may impede operations.

To ensure that situational awareness is shared successfully, partners should work towards sharing relevant information with all, give others an understanding of their perspective and priorities, their expectations of partners' capabilities and their own limitations, managing their partners expectations of them.

Effective situational awareness will therefore depend upon good lines of communication, the use of a common language (minimising jargon, acronyms, service or agency-specific language) and concise, precise reporting. It is important, however, that sharing information and situational awareness is a two-way process. Gathering information from others will ultimately improve all agencies' understanding. In the UK, the JESIP programme has enhanced information and situational awareness sharing through the introduction of joint briefing protocols and risk assessment methodologies, which enable a common understanding of incidents and risks, and of other organisations' capabilities and limitations. These protocols and the JESIP programme are examined more closely in Chapter 15.

Box 6.1: Unco-ordinated actions and miscommunications

A serious 10-pump fire in a riverside hotel was attended by operational crews from several fire and rescue services. The incident had been sectorised with crews from one service (Service A) undertaking firefighting activities on the ground floor in one sector and the other service's (Service B) crews working in the opposite sector using, among other things, an aerial ladder platform. During the incident a BA team from Service A took a jet into the premises to carry out firefighting on the ground floor at the rear of the building. The sector commander in charge of the aerial ladder platform directed a monitor onto the roof, which was on fire. The monitor jet struck a water tank in the roof space. The force of the jet caused damage to the supports and destabilised the tank, causing it to fall into the room below. The two firefighters fighting the fire were struck by the falling tank and one firefighter suffered severe leg injuries that led to his retirement.

Lessons learned

Incidents involving more than one service need careful management to ensure that firefighting activities are co-ordinated in order to avoid the possibility of dangerous situations arising from tactical decisions made in isolation.

Ensure incident ground communications are compatible with those of other services that may be required to provide reinforcement.

Joint exercising will help reduce the problems associated with multi-service working.

Adopting the principles of the Joint Emergency Services Interoperability Programme (JESIP) will assist in the process of information developing and sharing situational awareness. These principles are increasingly being adopted by a wide range of agencies and organisations such as railways, utility companies and airports, as they provide a common language for use during emergency situations.

Remote situational awareness

While situational awareness at an incident itself is vital, it is possible to overlook the potential ramifications of an incident at a wider, strategic level that is remote from the immediate location. Developing a remote situational awareness can allow those in a relatively isolated location to understand what is happening at the scene of an incident and anticipate alternative possible developments. This use of remote situational awareness can be particularly effective at large scale incidents where those at the sharp end may not be aware of events beyond their immediate location. The type of incidents where remote situational awareness is useful includes largescale wildfires, large area flooding or during multiple acts of terrorism across different parts of the country. A co-ordinated response to intelligence gathering, strategic prioritisation and the allocation of resources, sometimes on a national scale, will be required.

When those making decisions are remote from an incident, there is the potential for miscommunication. There is a naturally occurring time lapse between an event or activity taking place on the ground and the time that information reaches those based at the remote location which may render the received information redundant by the time it arrives. During dynamic events such as riots or civil disturbances, there may be a need to set up not only a strategic co-ordinating group but also a government level co-ordinating group (normally set up in the Cabinet office briefing room – COBR – see Chapter 16), and both these groups will require information to inform higher-level decision-making.

Box 6.2: Redundancy of essential information at larger incidents

During the Birmingham riots in 2011, requests for information by COBR were transmitted from the Cabinet Office via the Chief Fire and Rescue Advisers Unit (CFRAU) to the fire and rescue service incident support room. Officers in the support room then had to make contact with commanders in four different locations (the riots were spontaneous and simultaneous across the West Midlands, as was the case in London and other places), compile a report and transmit it to CFRAU, which would then feed the information to COBR. The time between the request for information from COBR to the receipt of the information required was between an hour and a half and two and half hours.

Lessons learned

This illustrates an important point: acting on information that is delayed may result in incorrect decisions being made or made too late. Wherever possible, information that is not contemporaneous (i.e. immediately available and current) should be time stamped at the time of origin and not at the time of transmission to make it clear to those who use that information that it may no longer be accurate or timely.

Loss of situational awareness

Situational awareness can be lost for a number of reasons and it is important for the IC to maintain a guard against this loss. Like many other processes, maintenance of situational awareness is continual and needs to be constantly checked to ensure that the IC's understanding of the situation is still current. As incidents move on, the dynamics of the fire or rescue etc. will change, the leak of hazardous substances may alter, even wind direction changes may have an influence on how a situation develops. It is therefore important that the IC continually checks their understanding of the situation and ensures that information is updated.

There are several indicators of a loss of situational awareness has occurred, and these need to be closely monitored. These include:

- a tendency to fixate or develop tunnel vision on one aspect of the incident

- a perception that an incident is progressing well despite no information confirming this belief – overconfidence

- being distracted

- suffering from information overload.

All of these indicators may show that the IC's perception of the situation may no longer be accurate and that a review of the information is necessary to confirm and/or modify the understanding of the situation.

Box 6.3: Failure to gain effective situational awareness

A terrorist explosion at a large event in a city centre led to the mobilisation of multi-agency resources to an operational rendezvous point. A Fire National Interservice Liaison Officer (NILO) was mobilised to manage the operational response at the scene. Suspecting that armed terrorists were present (but not having confirmed the details from a reliable source), the NILO independently changed the rendezvous point for the FRS responding appliances from less than 500m from the incident scene to another location 2km further away. There was in fact no armed terrorist threat (which was confirmed shortly into the incident). Due to the FRS vehicles being so remote from other services they were in effect isolated from the incident and were not able to deploy to the incident for two hours. The FRS were heavily criticised for their lack of activity at the incident.

Lessons learned

Do not take significant operational decisions when substantial information is lacking and you are remote from the scene of operations.

If accurate information is missing, then every attempt should be made to gain this information using other means: phones, radios, attending the police (or other organisation) incident control room.

Inform tactical and strategic commanders of difficulties and use their facilities and resources to acquire the necessary information.

Tunnel vision or fixation

A common problem for ICs arriving first at a scene is that they are faced with what appears to be a critical situation: there is a serious fire which needs to be extinguished or immediate rescues to be carried out and that becomes the focus of activities to the exclusion of all other, wider aspects of the incident – this is what we might call tunnel vision. For example, an aerial ladder platform is deployed to work as a water tower and the operator fails to notice high voltage cables within the operating envelope. Another common example is that of firefighting crews entering smoke-filled premises but failing to recognise the signs of an impending backdraft. If an incident commander maintains a focus on task-level activities such as pump operation or BA control, they may fail to observe an unusual growth of fire which may affect the structural stability of the building. This is another warning sign that situational awareness may have been lost due to tunnel vision.

All of these events have occurred and resulted in the serious injury of firefighters. Focusing on one aspect of an incident stops ICs from maintaining an overall view of the operations being undertaken.

There are a number of strategies that the fire and rescue services normally employ to help avoid tunnel vision. These include having the IC remaining on or close to the command unit once they have completed their initial information gathering – this allows them to view the whole of the incident rather than focusing or fixating on one part. Secondly, the safety officer or safety sector commander plays a key part in maintaining an up-to-date perspective on the incident and developments. Sector commanders, BA entry control officers and supervisors all play a part in updating the IC's understanding of the situation on the incident ground and should be encouraged to notify their respective commanders of any developing situations that have the potential to create additional risks to firefighters or other persons. The creation of an incident timeline grid (see Figure 6.1), setting out milestones including routine situation reports, analytical risk assessment reviews and updates, sending informative messages to control and updating sectors of tactical mode situation all help to ensure that information is regularly checked, reviewed and updated. This helps avoid circumstances whereby a focus on one aspect to the exclusion of others is allowed to continue and checked for extended periods of time.

Overconfidence

There have been circumstances in which firefighters have been seriously injured when they have believed that an incident is declining in intensity or has been controlled when in fact the situation is far worse. For example, during one apartment fire on the 16th floor of a high-rise building, firefighters who were in the process of damping down were engulfed by wind-driven fire when a window failed. Their colleagues who were outside the apartment at the time were in the process of making up equipment, confidently believing the incident had been dealt with and beginning to relax. On hearing calls for assistance, they re-entered the flat and rescued the two firefighters, one of whom subsequently succumbed to his injuries. There are also examples of firefighters being injured at road traffic collisions or car fires where gas struts have exploded as a result of electrical short-circuiting, sending projectiles up to 50m and causing severe shrapnel-like injuries. They had believed that the incident was virtually complete and the risk levels had diminished to very low.

It is possible to prepare for these types of incidents by reinforcing the training of firefighters so they 'expect the unexpected' and undertake 'what if' exercises. The IC, remaining in a location slightly remote from the 'sharp end', will have the physical view and the mental 'head room' to assess the wider risks and hazards that an incident can pose. This allows them to develop 'active foresight' to identify and act on risks that may not be immediately obvious to those working much closer to the action.

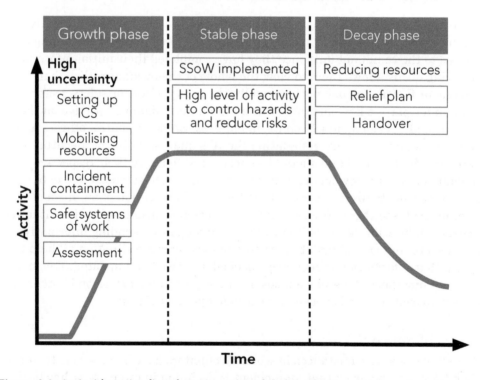

Figure 6.1: An incident timeline showing activity levels with time

Distraction

An incident ground is a busy place, with multiple simultaneous activities being undertaken and large numbers of communications networks and messages being transmitted often all at once. Even within the command unit there can be intense periods where information is coming from a variety of sources – communications with fire control, incident ground radio messages, meetings with other emergency responders and agencies, meetings with building occupiers, information being received from Mobile Data Terminals (MDTs) and briefings from tactical advisers and specialists. All of these sources could be considered as additional spans of control for the IC and have the potential to distract the IC or members of the command unit staff from 'seeing' the key aspects of the incident.

On occasion, ICs, while roaming the incident ground, have become embroiled in task level activities that have distracted them from managing the incident (see Box 6.4). Not only does this circumvent the natural chain of command, but it also takes the IC away from providing oversight of the incident. There are a number of strategies to reduce the possibility of this happening, including:

- Minimising the level of distraction: many command units have a conference facility or meeting area that can be relatively remote and isolated from the 'working end', which has incident ground radios, communications with fire control and meetings with crews and firefighters. By being physically separated from noise and distractions, ICs are able to focus more clearly on the incident having gained the mental space to think things through.

- ICs should be aware of the potential for distraction at operational incidents and engage the support of the command support officer and command unit team leaders, who can ensure that unnecessary disruptions to the IC are minimised.

- The comprehensive use of timelines and milestones provide regular checks on operational activities, messages, safety briefs and analytical risk assessment (ARA) updates, which will help keep the situation under review and help refocus activities if distractions have started to derail the ICs situational awareness.

Box 6.4: The 'Lost Commanders'

On a hot August day with temperatures reaching 25°C, crews were mobilised to a fire within the storage area of a large industrial premises. Twelve pumps and an aerial platform were mobilised and working under arduous conditions. Two level 4 commanders were in attendance: the Area Commander was IC and an Assistant Chief Fire Officer was monitoring the situation. A level 2 commander was the command support officer.

Both level 4 commanders left the command unit for a 360° reconnaissance in separate directions. While they were gone, several firefighters started to show the effects of severe heat exhaustion. This created a situation where a large number of firefighters were occupied providing casualty care facilities for firefighters. This created a shortage of resources to deal with incidents and additional resources were required. The transport officer was unable to contact either of the level 4 commanders and took the decision to request additional resources directly from control. The level 4 IC was eventually found inside the building directing firefighting teams operating jets, the assistant chief was discussing a sports event with a watch manager in charge of the water sector. Neither had radio communications with them.

Lessons learned

1. If ICs leave the command unit, they should always brief the person left in charge of the incident and make sure they remain in contact with a command support officer.

2. ICs should not be involved in operational activity on the incident ground.

3. The incident ground is not a place for social interaction – it provides distractions for others from key operational activities.

Information overload

As seen above, the increasing availability of various types of communications means that information overload is a real risk for the IC. Extracting key information from a storm of data can be difficult and is particularly difficult under time-pressured situations. The IC cannot afford to deal with all data that has been received and must rely on the command support team (and sector commanders etc.) to filter and prioritise the information that is required to manage the incident effectively. More experienced ICs will be able to carry out much of the filtering themselves, but it is always essential to let command unit staff, command support officers and sector officers know during the initial briefing what information is required from them.

Information and incident command requirements may be summarised as:

- information that the IC needs to know
- information the IC would like to be told
- information the IC does not need to know.

In this way, supporting staff and commanders are clear what level of information is required and the autonomy under which they operate within their part of the incident ground structure.

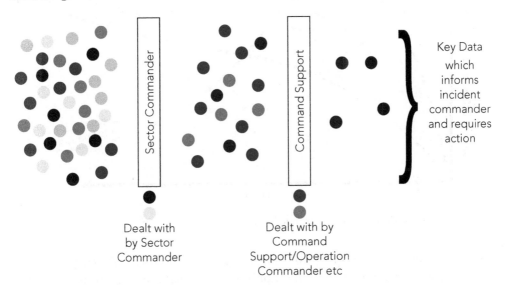

Key Data which informs incident commander and requires action

Sector Commander

Command Support

Dealt with by Sector Commander

Dealt with by Command Support/Operation Commander etc

Figure 6.2: Filtering out unrequired data

Information gathering

The key factor for maintaining a good level of situational awareness, effective decision-making and overall fire ground safety is to have a good level of information available from a variety of sources. Triangulation is the process of checking information from one source against that provided from another. Thus, initial mobilising information may be confirmed by radio traffic en route to an incident, observations, and by verbal reports from incident and sector commanders. If the details are consistent, it is likely that an understanding is more reliable and accurate.

Of course, it is possible that incorrect information is being reinforced due to 'group think' and confirmation bias may mean that all sources of information may be corrupted. This doesn't happen very often but the IC should seek to challenge assumptions made by others to ensure the robustness of the information being received. A good IC will be constantly looking for different sources of information which confirm (or challenge) their understanding of the situation, and this enables them to take decisions to influence the outcome of the incident.

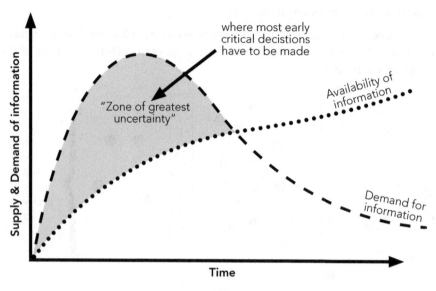

Figure 6.3: Information demand and availability

During an incident, the requirement for information and its availability will vary with the specific circumstances, the rapidity with which the incident is developing and the sources of information that are available. For most incidents, the demand for information is greatest during the initial (and usually most dynamic) phase of operations. Unfortunately, for most incidents, availability of accurate information is at its lowest during this phase and during the initial phase of an incident there is often a gap between the information available and the information required to make complete decisions.

This 'information gap' can never be closed completely and there may be 'unknown unknowns' that may lead to events that were entirely unpredictable. For example, the attachment of combustible ('limited combustibility'!) weather screens to the outer faces of high-rise buildings were not considered to be inappropriate (or illegal) before 14 June 2017 and so firefighting tactics and operations did not take into account the possibility of a high-rise fire simultaneously burning over 20 or more floors.

For most 'normal' incidents, a catastrophic lack of information is not normally the case but decisions still have to be made when insufficient information is available. This gap, sometimes known as the 'zone of greatest uncertainty', means that the IC will be making decisions in the absence of full information (see Figure 6.3). ICs need to accept that 100% of the knowledge required to make a 'correct decision' will usually be an unattainable goal. What is needed is an acceptance that as much information as available should be used to make the best decision possible at that particular time.

Colin Powell, former US Secretary of State and four-star general, devised a guide for making balanced judgements when decisive action is needed but information is lacking – the '40% / 70% rule'. Essentially, this rule states that gathering between 40% and 70% of available facts and data and then making a judgement based on your knowledge, experience and 'gut feeling' is an acceptable calculated risk, while accepting that information is incomplete. Where less than 40% of the facts are available, then a decision becomes a high-risk gamble with a potentially negative outcome.

The time it takes to collect data beyond 70% can mean that the opportunity for making an effective decision is lost and events have moved on, with the worst outcome being no decision taken or being so delayed as to be useless. It is important however, that ICs who tend to make quick decisions should always ensure that they consider all options before dropping into what could be termed 'recognition prime decision-making' mode, being overly reliant on 'gut feelings' and less reliant on facts (see Chapter 7).

Sources of information

As we have already seen, the greater the number of sources of accurate information, the more likely it is that any operational decisions and plans will be based on firm foundations. This information will come from a number of sources and include the IC's own knowledge. There are wide variety of other information sources which will help furnish the IC with the particular details of an incident and which will help them finesse their understanding of what has happened, what is happening and why, and what is likely to happen in the future. These sources of information are discussed in greater detail below.

Before the incident

Many premises will have site-specific information available that can be used by the IC and accessed before or during an incident. Site-specific risk information (SSRIs) can be of varied formats, such as that set out in chapter 3.

Most common premises will have generic risks that will be well-known and the IC will be well prepared for. Knowledge of an area will help inform the IC about certain types of building (Victorian, Edwardian or modern construction, for example) and the occupancy of those premises and their associated risks and hazard levels. For high-risk premises, there may be other statutory plans including those for the Control of Major Accident Hazard Regulations (2015) (COMAH) which the IC should be notified about upon arrival, as should each appliance and officer deployed to the incident. All appliances should have access to this information via mobile data terminals (MDTs) within the larger appliances, although firefighters and the IC should be aware of the existence of these premises in their geographical area of responsibility.

At the turnout

Turnout information will invariably comprise a teleprinter message giving basic details such as:

- the type of incident
- the type and number of appliances attending
- incident address
- type of premises
- additional information sources including SSRI, COMAH and generic risk information.

It may also include information about the route to be taken, particularly in the event of the hazardous materials incidents or where access may be restricted or problematic such as motorways, railway infrastructure or airports etc.

For levels 2, 3 and 4 ICs being mobilised from the office or home environment, a pager message may give a basic level of information about an incident but there is an opportunity to gain additional information from the mobilising control (by a follow up phone call) as it is likely that fire and rescue resources are already in attendance having been mobilised following the initial call. By careful questioning of mobilising control, it may be possible to start the development of an added layer of situational awareness before the IC begins their journey to the incident. Information that may be useful includes:

- The additional resources that are being mobilised at the time.
- The identity and organisational role of the current IC, which will help the potential IC gain an idea of the level of experience and knowledge that individual may have.
- The time of the call. This may help develop an understanding of the fire dynamics – an early mobilisation of additional resources may be indicative of a fire or incident that is already well developed or escalating.

- The time of receipt of assistance messages. This information may help determine the speed at which an incident is developing. For example, the request for an additional two pumps at 20 minute intervals (i.e. 'Make pumps 4' at 08:00 hours, followed by 'make pumps 6' at 08:20 hours and 'make pumps 8' at 08:40 hours) may indicate that the fire is getting more severe at a slow pace or that the IC is reluctant to request a large number of reinforcing appliances in one message. Conversely, a rapid and early make-up is likely to indicate that an incident is developing quickly and may have the potential to grow even further. Informative messages will assist the commander in understanding the tactical mode at the incident and possibly the reason for the selection of that tactical mode.

- The number of resources in attendance and those en route to the incident will help the IC assess current progress being made against plans for their arrival with a view to assessing what future requirements may be.

- What additional officers and specialist support personnel are attending the incident, including hazmat officers, fire safety officers, safety sector commanders, transport officers etc. This information may help the IC plan a possible incident command structure before they arrive at the incident.

- Any control measures that may already be in place. This includes the evacuation of residents from the affected building or adjacent premises, the stopping of railways or major roads and even the implementation of air space restrictions. This information may be available from mobilising control as a result of follow-up emergency calls and informative messages from the incident.

- As the seniority of the mobilised officer increases, availability of additional resources will become a concern. It may be relevant to ask if there are any other incidents taking place within the service area that may have an impact on the speed and weight of reinforcements as well as the availability of special appliances that may be required, such as aerial appliances, water tankers, rescue tenders, high volume pumping units etc.

En route

While travelling to an incident, mobile phones (hands-free) and radios should be monitored to maintain a general awareness of activities at the incident being attended and to listen to any messages been transmitted from the current IC. It is important, however, that priority is given to driving safely and giving full attention to maintaining control of the vehicle.

On arrival at the incident

There are a number of sources of information that will support the IC gaining all-round insight into what is happening.

The initial (and subsequent) IC as a source of information

The initial IC, upon arrival, will be faced with a number of competing priorities including the carrying out of rescues, attacking the fire and stabilising the incident. Simultaneously they will need to continue the process of gathering information, which started following the first alert. The sources of information available to the first IC can often be very limited and consist of what can be seen, smelled and heard by themselves. As more senior commanders arrive, their level of information should be greater and support a more detailed briefing to the oncoming commander.

The 360°

The visual cues available are one of the primary methods by which the IC can assess an incident. The fire itself, exposures, wind direction, building height and dimensions, all help the understanding and perception of the situation. To get the best understanding of the layout of the incident, a walk round the incident ground itself, i.e. a 360° reconnaissance, should be carried out, ideally by the IC, or if impracticable, a nominee. This will enable the IC to get a 'feel' for the incident ground and comprehend events in areas beyond the immediate view of the command point or unit, having already reconnoitred each sector.

There is an opinion that more senior commanders attending the incident need not reconnoitre the incident ground for themselves but react on the basis of information gathered by others. Although this may be a matter of personal preference, it would appear at the moment that most senior commanders still prefer a personal reconnaissance of the incident ground wherever practicable. For larger incidents, technology now allows remote cameras to be positioned in locations remote from the command unit, allowing the IC to see activities within specific sectors. Unmanned Aerial Vehicles (UAVs) or 'drones' are becoming available to most services and where possible should be used to support the IC in developing a complete and live view of the incident. Once again, while the technology works (and doesn't get stolen) a good understanding of the incident ground should be obtained. However, ICs should always be prepared to carry out a personal reconnaissance to gain that valuable incident ground knowledge.

Eye witnesses

Eyewitnesses can be useful but may have a limited understanding of what has happened and what is happening. A lay witness, without specialist knowledge or experience of fire, should not be relied on to give authoritative information about an incident, but they should be able to give some factual details of what happened and what is happening. Examples of an IC being led astray by a lay witness include incorrect reports of persons being in a building on fire and the witness surmising what happened before their arrival and presenting their assumptions as facts. Nevertheless, the IC should retain the presence of potentially useful witnesses and use command support unit staff to gain information from them.

Premises' owners and occupiers

The occupiers or owners of buildings involved in fires may often have specialist knowledge that will enable the IC to gain useful knowledge of risks, fire safety measures and construction etc, which may assist in dealing with the incident. Occupiers should also be able to identify locations and specific hazards associated with processes, storage and utilities. Sometimes they may also have access to plans specifically for use by the fire and rescue service (which may also sometimes be found in specially designed, FRS accessible security boxes on the outside of many industrial and commercial buildings). Premises security guards are also likely to have a wide range of information and can be of use in leading crews to difficult-to-access parts of buildings across complex industrial sites.

It is important to remember that the IC has a duty to protect the health, safety and welfare of occupiers and eyewitnesses at an incident and should be mindful of the need to protect them from harm that may arise from fire and rescue service activities.

Site Specialist Advisors

At some sites, due to the complexity or risk of work activities being undertaken, there may be specialist advisers who are able to support emergency services during incidents. The specialists may be workers on site with specific knowledge of hazards and control measures who will be able to advise the fire and rescue service of actions to be taken. In most circumstances this should be of great benefit to the IC, especially when this advice has been developed with the support of FRS specialist staff such as a hazmat officer. Where a decision for action has been made based upon the advice of specialist advisers, the decision should be logged and the rationale for the decision should include the information given by the adviser. Other types of on-site advisers include environmental protection advisers and radiation protection advisers for premises where radiation is suspected.

Remote sources of information

The widespread availability and reliability of mobile communications technology means that the need for carrying vast quantities of paper-based information systems is no longer necessary. Mobile data terminals (MDTs), computer-based command support systems, digital cameras and mobile phones may provide an IC with visual as well as written information about an incident. For example, real time imagery of an incident is now possible via a downlink from police or military helicopters or the FRS's own UAVs, which can support and monitor incident progress directly at the scene. Services' own risk databases can be immediately accessed giving SSRI information and plans on-site and, if necessary, search engines can assist the immediate acquisition of information about a premises. These systems may allow the IC to share information with members of the command team, colleagues from other agencies and personnel/commanders remote from the incident.

There are some important issues to remember when using remote sources of information. When visual information is communicated in such a manner it is important for the receiver to understand the context surrounding it to avoid misunderstanding. For example, the meaning and significance of symbols on a plan, or the implications of the properties of materials involved in a fire need to be understood when determining the operational tactics available. Having visual and other electronic information does not mean that other forms of communication are no longer required. Even if an IC has sent pictures of the incident via an email to a remote location, it does not mean that a shared understanding of the situational awareness has been achieved. Such information must be supported by verbal or written communication to ensure the context and message an IC wishes to convey is understood.

It is also important that the security of the IT networks over which such information may be transmitted and received is secure in order to protect against sensitive information from reaching unauthorised individuals, including the media.

Chemical meteorology (CHEMET) service at the Met Office

In the event of an incident involving hazardous chemicals, such as a spillage or fire, an RTC involving hazardous materials spilled or on fire, local police and fire services contact the Meteorology Office (the 'Met Office'). There is a need to understand the likely spread of a plume or fumes into the atmosphere in order to anticipate actions that may need to be taken to control it, to determine which areas may need to be evacuated and to assist with recovery and remediation processes. The Met Office use computers to factor in precipitation levels, topography, localised wind speed and local variations, the stability of the air mass and the impact of changing temperatures.

The report produced is known as a CHEMET and consists of two parts – a text forecast and an ordnance survey map with an overlay which shows the areas most at risk. The CHEMET forecast may be used by the IC or hazmat officer to assist the preparation of plans that anticipate the likely consequences of a continuing incident. Producing a CHEMET report can take some time, so a more rapid system, FIREMET, also from the Met Office, provides immediate access to a forecast of conditions. While less accurate than the CHEMET, it can assist first responders with data that forms a dynamic and analytical risk assessment and inform the preliminary deployment of safety cordons.

Other remote sources of information

Other sources of information may include:

- Detection and Investigation Monitoring (DIM)Teams.
- Specialist Managers e.g HMEPO.
- Specialist teams e.g high volume pumping units, urban search and rescue teams.
- The Environment Agency, Scottish Environmental Protection Agency, WELSH and IRIS versions.
- Network Rail and train operating companies.
- The Civil Aviation Authority and airports authorities.
- Military resources, such as explosive ordnance teams who can provide information and advice on unexploded weapons and ordnance.
- Health boards and Public Health England.
- Local authorities.
- Utility companies.
- The internet.
- FIREMET.

This list is not exhaustive, but it indicates the variety of advice that can be obtained remotely. Ideally, these sources of advice and information would be identified and contacted by Fire Control, incident rooms (where set up) or through the command unit in the first instant. It is not the role of the IC to carry out research of this type – it is a function to be delegated, and only the outcomes need to be communicated to the IC. It is therefore important that when delegating the task on information gathering to others, the IC gives a clear briefing on what outcome is being sought and should ensure that the individual to whom the task has been delegated is clear about the parameters of the task.

Subsequent command changes

Officers who attend incidents following the initial attendance will have the advantage of having 'boots, eyes and ears on the ground' before they arrive, and so much of the initial gathering of information will have been acquired by the initial IC. Unless there is an urgent need to immediately take command of the incident, higher level commanders have the luxury of time to gather details of the incident, form ideas, strategies and tactics before formally taking command. Where an incident is being managed effectively, it may be possible to undertake a personal 360° reconnaissance to get an orientation of the incident and a flavour of current operations. This will help triangulate information already received from mobilising control, messages on the radio and pre-existing knowledge. The final piece in the information gathering phase is the current IC who should be in a good position to brief the oncoming commander.

Any briefing of an oncoming commander should be comprehensive and complete so that the new commander is fully conversant with the incident, the IC's plans and their intentions for future operational activities. There are many different systems that are used for briefing, including the use of the decision control process, mnemonics such as OTHERS (see the example below) or a basic description of the incident, actions taken and intentions, sometimes supported by the use of messages sent to fire control. Whatever system is used, the information passed must be accurate, unambiguous, current and comprehensive, but also brief and delivered clearly.

Such briefings should always be supplemented with any messages sent to fire control, a review and handover of the Analytical Risk Assessment, the safety briefing, the SSRI and any specific risk plans (e.g. COMAH or off-site plans), confirmation of tactical mode and, importantly, a review of the decision log and the rationale behind those decisions.

OTHERS briefing

This is a mnemonic in common use in the UK for briefing officers who are considering taking command of more serious incidents:

O   Operational objectives

T   Tactical mode

H   Hazards identified

E   Existing deployment

R   Resources available

S   Structure of communications and messages sent

Figure Box 6.5 provides a detailed example of a handover using the OTHERS mnemonic for a factory fire involving chemicals.

Box 6.5: Example of an OTHERS briefing for a serious fire involving chemicals

- **Operational objectives**
  - Rescue emergency team members in building.
  - Contain fire to bay 3.
  - Contain run off within site.
  - Liaise with police and Environment Agency (and PHE) regarding wider implications of incident.
- **Tactical mode**
  - OFFENSIVE ALL SECTORS.
- **Hazards identified**
  - Severe fire in bay 3.
  - Calcium Hypochlorite and Trichloroisocyanuric acid especially if involved in fire.
  - Rapid collapse of bays 3 and 4 leading to large smoke plume.
  - Exposure of public to smoke and fumes from fire.
- **Existing deployment**
  - First 6 pumps clogged up inside the premises and in the access road/junction of Napier Rd.
  - The CSU and remaining 7 pumps at Sussex Exchange and Queensway.
  - Sector 1 – 6BA, 2 ground monitors, 1 jet, 1 HRJ – search teams of 2 and 4 deployed in Bay 3/Bay 3 (WM Sector Comm).
  - Sector 4: 4BA, 2 Jets (WM Sector Comm)
  - Initial decontamination set up in sector 2 (near office block) – we have cut down the perimeter fence and made a path that leads from clean area to Napier Road (WM Sector Comm).
  - Water Sector and Marshalling set up.
  - Command structure – IC, Ops Comm, CSO, Safety, HMEPA
- Resources available
  - 13 Pumps in attendance.
  - 5 full crews awaiting deployment from holding area as reliefs for BA.
  - 2 Pumps due in to make full attendance complete (MP 15).
  - Police, ambulance in attendance; Environment Agency eta 15 minutes.
  - The works manager is in attendance.

- Structure of communications and messages sent
  - Fireground 1 Command channel to sectors.
  - Fireground 6 for BA.
  - Fireground 4 for water and support sectors.
  - Informative and assistance messages sent at ....

It is equally important that a systematic approach is taken when handing down an incident to a more junior commander. A commander's desire to leave an incident that he has been attending for a number of hours can be strong, but the handover is a vital component in maintaining a safe working environment and it ensures that safe systems of work are identified and their purpose explained. A superficial and cursory handover is dangerous as it may leave the impression that the incident is being downscaled and risks have diminished. There have been many incidents where the speed of the handover has been such that key safety information has been missed and, as a result, firefighters have been injured or killed. Handing down processes should match those when handing up and time should be taken to show the relieving teams and officers the roles and activities that those being relieved were carrying out. Again, safety briefs, ARAs and orientation of the incident ground should take place at incident, sector and team levels before the relieved depart the incident ground.

Failure to gain information

Where information is incorrect, absent or incomplete there is an increased risk to the safety of those on the incident ground and could allow unexpected developments to occur without any anticipation. Some of the consequences of a lack of accurate and timely information include:

- The development of a narrow perspective of the incident – focusing on part of the incident about which information is known and missing others due to the lack of information – the 'unknown unknowns'.

- Confirmation bias, where information that supports a commander's understanding is accepted but information that contradicts that view is ignored as an 'uncomfortable truth'.

- The lack of knowledge of risks at the incident may cause firefighters to be inadvertently exposed to dangers that could easily be mitigated if identified.

- Unforeseen consequences of actions taken in the belief that knowledge is complete and correct, where the actions in fact exacerbate the risks to firefighters and the public.

Cognitive dissonance and that 'gut feeling'

It is possible that on occasions, the IC in waiting will experience what is termed 'cognitive dissonance'. This can occur in many aspects of day-to-day life but on the fireground it can be described as the uneasy feeling that when 'things just don't add up'. For example, messages from fire control may give the impression that an incident is being controlled and moving towards closure, and on arrival at the incident the IC confirms that the incident is being controlled and will shortly be downscaled, confirming the initial perception from control. Carrying out a 360°, the IC in waiting finds that a deep seated fire is behaving in a manner that suggests the fire is not under control and resources may need upscaling. The discomfort and tension that may be felt because of the two conflicting messages (fire under control: fire not under control) is an example of cognitive dissonance. Sometimes this can have obvious reasons such as the example above, but it can equally be the case that at times it may just be an uneasy feeling that things aren't going to plan. Don't ignore gut feelings: they are often the product of years of experience, memories and incidents previously attended and are a valid manifestation of real concerns.

Filtering out the chaff – the role of the command support staff

The role of the command support team in filtering out unnecessary or irrelevant information or detail ('white noise') is vital. In terms of information, the incident ground is a busy place and the IC will be bombarded with data from a range of sources. While a lack of information can be a risk critical failure, too much information can also result in heightened levels of risk to firefighters. Incident ground, radio noise, mobile phone communications, internet and social media, face-to-face meetings with FRS staff, other blue light services, agencies, members of the public, can all threaten to overwhelm the IC and result in key risk-critical information being lost. An experienced command support team will help organise communication flows and help filter out extraneous information, leaving the IC with the headline issues and key details to enable decision making to be undertaken quickly without having to trawl through details. The command support officer and command unit supervisor will also help control the flow of information from visitors and agency personnel, capturing key information and allowing the IC to get on and command the incident.

Summary

You should now understand the following concepts, techniques and practicalities discussed in this chapter:

How situational awareness originated and its importance for incident ground safety, and hence the role of the incident commander.

How sharing situational awareness across agencies ensures the best outcome for responders and the community and how loss of SA can impact on their safety.

How tunnel vision and decision traps may lead to firefighters' safety being compromised and lead to an unsatisfactory conclusion to the incident.

How information gathering from a wide range of sources and the use of filtering aids situational awareness.

The use of the OTHERS method of briefing provides a comprehensive overview and details of an incident for oncoming commanders and crews.

The appreciation of 'gut feeling' when seeking an opinion of progress of an incident.

Chapter 7: Decision making and the development of the plan

Introduction

After gathering the relevant information and developing a situational awareness, the next phase is to move into decision-making, and devising and implementing a plan of action. A decision is the conclusion or resolution reached after consideration of the pertinent facts, circumstances and options. It should lead to an action that may be positive or negative. Most fire and rescue service decisions are not made on the fire ground but rather in the less critical environment of stations and offices. It is, however, the decisions made at the incident ground that have the potential for causing immediate harm or good to both members of the public and firefighters. Decision-making in pressurised circumstances has been the subject of much research and a range of decision-making processes and methods are widely used across industry, government and the emergency services. It is not the intention of this chapter to discuss the academic psychological ideas underpinning the concepts and theories of decision-making in emergency services, but rather to give an overview of how they can be used at an incident.

The first decision: taking command

After gathering sufficient information about the incident, the oncoming IC will need to assess whether to take command or not. In making this decision she/he will need to provide themselves with a rationale for doing so. The reason may be that the scale of the incident, number of resources attending, or FRS policy require they take over. As this is a formal step in the incident command process, the change of command should be formally stated to the current incident commander using the phrase *'I am taking command'* so that no-one is in doubt of the change.

The current incident commander should then be given a suitable role that helps continuity of the incident and provides support to the new incident commander.

Decision making

The way in which we come to a decision about what action should be taken will depend on a number of variables. These include the level of certainty about the situation we are faced with, the time we have in which to make the decision, our experiences of similar situations in the past and our knowledge, skills and expertise at managing events. Extensive work has been undertaken by academics and practitioners in the field of decision-making: ICs invariably operate in an environment where time is limited, information restricted, goals are significant and decisions need to be made very quickly. The process of making a decision is part of an iterative process where information is gathered and assessed, options for actions are identified, and the best option is selected and then implemented. This is then followed by a revaluation of information and a modification of decisions and actions etc.

The type of decision-making process that is undertaken will depend upon a number of factors including the experience of the IC, the time available for that decision (which may depend on a wide range of considerations) and the risks to both firefighters and casualties of any actions undertaken. It is important that the IC understands the various types of decision-making processes so that they have a range of 'tools' that can be used appropriately across a range of circumstances.

The incident management process

As discussed in Chapter 3, incident command is a systematic and iterative process with a set of steps and sub processes that generally, but not always, flow in sequence, sometimes carried out instinctively, sometimes in a more deliberate manner, with inputs, outputs and resolutions achieved. This process (expanded from the list in Chapter 3) follows on page 99.

Situation assessment

An understanding of the situation that the IC faces will be developed from a wide range of sources, including the information available prior to attendance at an incident, prior knowledge of the premises or type of event that has been attended and by gathering additional cues and details upon arrival at the incident, all of which will combine to create an initial level of situational awareness (see Chapter 6).

Information gathering.

↓

An assessment of the incident or situation – deciding what the problem is.

↓

Assessing the risks.

↓

Determining the strategic objectives and tactical priorities.

↓

Identifying tactical options.

↓

Selecting an option.

↓

Setting an outline of tactics to match the needs of the strategic and tactical priorities.

↓

Assessing the resources required based on those priorities.

↓

Implementation of tactics to control, contain and neutralise the hazards.

↓

Organise the incident: structuring the incident, sectors, sector tasks, allocation of resources, communications structure, administration (safety reporting, messages, reliefs, briefings, reporting schedules and updates/sitreps).

↓

Taking steps to mitigate the impact of the incident.

↓

Closing down the incident.

↓

To help undertake the post-incident review and identify lessons learnt in order to improve the management of future incidents.

Types of decision making

The decisions we make can very often depend on the circumstances occurring at that precise moment in time. Decisions made on the incident ground can be influenced by a wide range of factors including our mood at that moment, which may be influenced by personal, domestic factors, our energy levels, time of day etc. Experience of previous incidents, training and a good appreciation of the situation being faced will all help in making an effective decision. There are a number of ways in which decisions are made. Some are quite instinctive and others more deliberate, requiring consideration and analysis. The type of decision making process we undertake will depend on the situation (if lives are at immediate risk for example, or if the incident is developing rapidly, etc.) and the amount of time we have to make those decisions. There are a number of models that are commonly used by ICs and are briefly explained below.

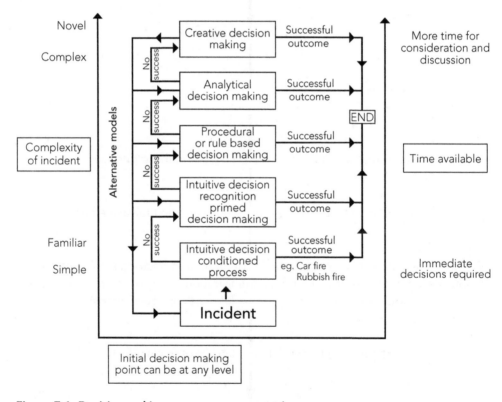

Figure 7.1: Decision making processes – an overview

Intuitive decision making

Intuitive decision making is an instinctive process using what is termed 'a non-sequential information processing load' (Sinclair, 2005) and uses information gained through previous learning that is stored in the long-term memory and accessed unconsciously. In many respects, this is similar to the use of 'gut feeling' that we rely on to make many decisions in everyday life. For the experienced IC, many incidents will have a high degree of commonality which means that arriving at a bedroom fire in a semi-detached house will be relatively familiar and the 'experienced' IC will act almost reflexively and instruct crews on the basis that this incident type is almost generic. Indeed, most firefighters with a degree of experience will also undertake tasks automatically and it is likely the breathing apparatus sets will be started up, firefighting media organised and action started within seconds of arriving at an incident. In many cases, the IC will have reacted to the visual cues, made decisions without planning systematically (or even consciously), and automatically set in motion actions to deal with the situation.

These types of reaction can either be through conditioned processes (CP) or recognition prime decision making (RPD). Let's have a look at both of these now.

Conditioned processes

Conditioned responses are those that become automatic following training that 'conditions' individuals to react to stimuli like Pavlov's dog experiments. Infantry soldiers are conditioned to react instinctively to certain words – for example 'CONTACT LEFT' (meaning an enemy has fired a shot from the left side of the axis of advance) will result in a conditioned response with them immediately going to ground and taking actions to neutralise the source of threat. An example for the FRS might be the automatic response of a firefighter to a loud 'STILL', an order given to cause an immediate stop to an activity which has become extremely risky. It has been ingrained in firefighters attending training centre from day one. Sometimes a breathing apparatus wearer upon hearing the pre-alert of an automatic distress signal unit (ADSU) will instinctively shake their shoulders to stop the pre-alert.

Because these processes are automatic, they are quick and very reactive. One of the advantages of conditioned responses are that they can be a protective mechanism, protecting the individual from harm, such as ducking down or blinking at the sound of a loud noise or explosion. On the other hand, they can be a disadvantage because the responses become instinctive and therefore they can become automatic even in inappropriate situations. It may mean, for incident, that an individual may expose themselves to intolerable levels of risk because their instinct is to act.

Consider, for example, the automatic response of a firefighter (or police officer or paramedic) to the sight of a person or animal in difficulty in a river or sea. There have been many incidents in which potential rescuers have lost their lives in circumstances that were extremely risky and where a rational decision based on a risk assessment would have determined that activities other than direct intervention may have achieved the desired outcome without creating additional risk. These very human reactions are instinctive and undertaken with the best of intentions but can often end in tragedy. Training in the use of the STAR model of personal risk assessment, dynamic risk assessment processes and effective fire ground management can help reduce the likelihood of condition processes being undertaken at operational incidents.

Recognition primed decision-making (RPD)

Naturalistic decision-making (NDM), originally developed by Gary Klein, is the study of how people make decisions and undertake complex functions within challenging, real-world contexts that are high-pressure, high-stakes, time limited with high degree of uncertainty, situational instability and varying levels of experience and knowledge. Recognition primed decision-making (RPD) is the main model, but this was derived from Klein's NDM work.

RPD is a reflexive decision-making process that enables ICs to make quick and effective decisions when faced with complex incidents. Individuals use their experience at the current event by matching patterns from previous events using cues, anticipated outcomes, goals and actions. By doing this, decisions are made quickly and successfully, and this process can explain why some people can make good decisions without going through the process of developing and comparing options.

Klein devised three levels RPD model, which varied from a 'simple match' (Level 1 RPD) where the IC assesses 'four aspects' of the incident – plausible goals, relevant cues, expectancies, and actions – and matches the current situation to previous experiences. They then implement actions based on those experiences. The Level 2 RPD model ('diagnose the situation') involves developing a further course of action that requires the creation of a mental simulation of the impact of actions being taken. When it is felt that this action will work, then implementation occurs. Where it may work with minor changes, then the actions should be modified. The more complex the situation, the more considered the strategy. The third model, Level 3 RPD, can be used where the decision maker has less confidence in the situation and carries out a brief mental simulation of the course of action to identify any potential problems. If there are problems, the course of action is then modified and the mental simulation is replayed.

In experiments, ICs have been found to make mental simulations of how an incident would develop from the current situation. Where ICs believed that planned actions would be successful, a plan of action was be implemented. Where an IC believed a course of action *could* work, the IC would try to adapt the solution or consider a different method to achieve the desired outcome. This pragmatic approach values speed to implement and achieve over delivering the best possible solution, which may take some time to identify, assess and implement.

This is logical because at fires and other emergency operations, the situation is dynamic and very often a fast but imperfect course of action may be better than waiting to find a better solution as the circumstances and incident may have deteriorated. According to Klein (1989), his research with fire commanders who had faced challenging incidents indicated that between 80% and 90% used RPD-based strategies to resolve incidents. Other research supported these findings, and found that actions taken as a result of an RPD approach were much better than actions taken based upon other methods.

Apart from the speed of decision-making there are also a number of other advantages to using RPD, including the fact that little conscious thought is needed. It usually provides a workable solution when used in 'routine' situations and is reasonably resistant to stress (Flin *et al*, 2008).

The downside of RPD is that there can be a tendency to shoehorn an incident into a solution rather than modify an existing solution to meet the needs of the incident, with ICs looking for supportive evidence (confirmation bias) rather than looking for cues that would challenge their presumptions. It is also necessary for the IC to be sufficiently experienced and knowledgeable of similar situations, because without prior experience it will be difficult to make professional judgements. There is the potential for falling into the mistaken belief that all incidents follow similar patterns and should therefore be treated the same. For example, how often has it been the case that the first attending IC on a pump instructs the deployment of two breathing apparatus wearers to tackle a fire with a hose reel jet, even though the incident is a fire in a warehouse that eventually ends up requiring 10 main jets to control?

Another problem with RPD is that because some types of incident are seen repeatedly, such as kitchen fires, bedroom fires, and even in high-rise buildings there may be a fairly relaxed approach to these incidents, which are sometimes regarded as 'bread and butter jobs'. Unfortunately, this type of thought process has led to firefighter deaths due to under estimation of the situation, believing that the fires they were dealing with were similar to those in their past experiences and not taking into account differences in the current situation.

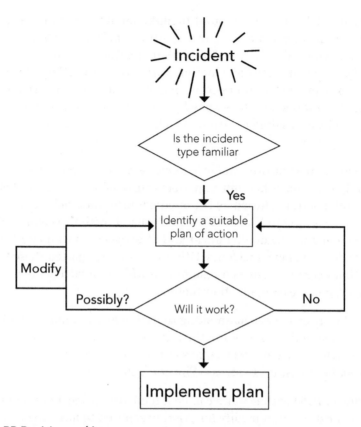

Figure 7.2: RP Decision making

Procedural or rule-based decision making

Fire and rescue services have policies, procedures and protocols that cover many types of incidents and situations. It is likely that for each incident type there is a specific procedure for which there are documents, accessible in paper or electronic forms, which give guidance (and on many occasions mandate instructions that must be followed) as to how to resolve a situation. An experienced and competent IC will be aware of the existence of SOPS (standard operating procedures) and be familiar with the essential issues if not the detail. Procedures for incident types that are less common or more complex than those judged to be 'routine' will require a conscious effort to recall, or will need to accessed from physical or electronic sources. Clearly, at a dynamic and fast-moving operational incident, the actions of firefighters should not be unduly delayed while procedural information is retrieved. This is because acquiring rule-based decision-making knowledge using operational procedures is usually part of the development programme for

ICs at all levels. Memorising key procedures will allow the IC to evaluate and consider possible actions relatively quickly, but the more complex the incident then the more consideration needs to take place.

The advantage of procedural decision-making is that it can be used by relatively inexperienced commanders while they are still learning, using experiences of those who have used the procedures (experts) in the past. With experience, implementation of procedures can be relatively quick and, because the procedures are used organisation-wide, they will be understood by those who have to implement them, i.e. firefighting teams.

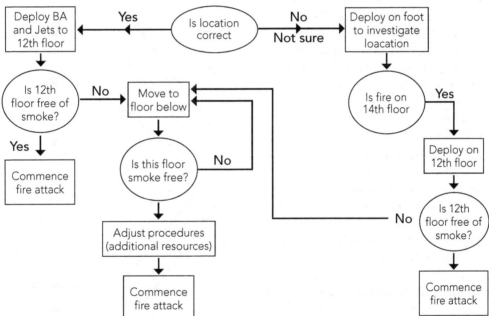

Figure 7.3: Procedural decision making

Where the procedures are not readily known or accessible, then using this process can take some time, which may not be available at many incidents. This may create stress for both the IC and firefighters as the desire to take action is held back by the necessity of determining the correct procedures that should take place. Such procedures should be reviewed on a regular basis to ensure they are kept up-to-date and also practised: relying on a procedural document during an incident is fraught with difficulty (access to materials, time pressures, unfamiliarity with the document) and has the potential to increase the risk to firefighters.

It is also the case that wrong rules may be selected, which can result in the wrong procedures being followed or, as has happened in the past, procedures being used that are not a precise fit for the circumstances, resulting in inaction which has caused additional damage and increased the risk to firefighters and members of the public. The importance of continually reminding staff of operational policies and procedures cannot be over-emphasised – a fundamental understanding of procedures is essential to securing beneficial outcomes and minimising risk.

Analytical decision making

Analytical decision-making tends to be used at more complex incidents where rough and ready solutions or existing procedures are not appropriate. This process follows a number of steps, which includes an analysis of the problem, consideration of existing rules and prior experience, the generation of potential solutions (which may be pre-existing or original), and the selection of the optimal solution. By necessity, and despite creating shortcuts where potential solutions are considered in the light of experience, this type of process can take some time and it is not likely to be undertaken in the growth stage of an incident. Rather, it is more likely to be undertaken when the incident has developed into a relatively stable condition or steady-state (such as would occur following the evacuation of the public from a hazard zone of

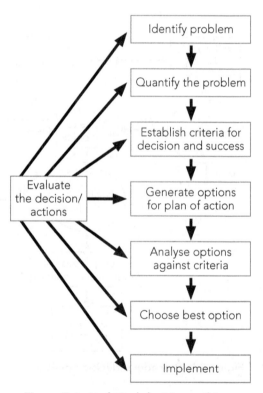

Figure 7.4: Analytical decision making

the hazardous materials spillage, following the extrication of the casualties from a vehicle where recovery of the vehicle remains a hazard).

The advantages of analytical decision-making are that it is methodical, robust and compares alternative solutions. It is more likely to arrive at an optimum solution than many other decision-making processes as time is taken to consider alternatives. Where possible, analysis of equipment, techniques and personal skills will factor into this process and will form part of the rationale for the decisions made. By necessity, this process is not quick and it requires an environment that is suitable for discussing and debating options, which is something not always available on the incident ground.

Creative decision making

Creative decision-making is something that should generally be avoided in high-risk, time-critical situations: it should only be used when solutions cannot be found through existing procedures or modifications of those procedures. This style of decision-making can be used when faced with an unfamiliar problem that may require a new solution. As the solution is likely to be pushing the boundaries of what would normally be required, it may be both time-consuming and untested. The balance of risks versus benefits must be rigorously assessed, particularly if the solution could result in catastrophic consequences in the event of failure. The use of a tower crane, for example, to rescue a casualty with severe back injuries from a high-rise building under construction, rather than more conventional means of rescue (due to access restrictions for an aerial ladder platform and medical urgency). It should be noted, however, that in the above real-life situation, alternative methods were available but for reasons which were valid at the time, this solution was adopted.

Creative solutions and operational discretion

There is a fuzzy boundary between what constitutes a creative solution and the use of the doctrine of operational discretion. The creative element in finding a solution to an unfamiliar situation can result in a variation of existing procedures, the unusual use of FRS equipment or the use of a 'created' solution to achieve an outcome. This variation is called 'operational discretion', which will be considered in greater detail in Chapter 11. Creative solutions and decisions inevitably take time to develop as there is a certain amount of uncertainty as to the risks and hazards that may be created by, for example using a piece of equipment for purposes for which it wasn't designed. This could be the use of breathing apparatus sets for carrying out a rescue underwater: something that FRS BA was not designed or intended for. The example of using working from height equipment for a rescue in a disused mine is a creative method of working but is also operational discretion. Again, creative decision making takes time, something that can be very limited in an emergency.

Decision support tools

Operational incident decision support tools are designed to help the IC structure their thoughts systematically and rapidly, and to allow decisions to be made based upon:

- the knowledge of the situation
- the inherent and imposed risks
- the resources required
- the development of a tactical plan that aims to neutralise the threat, rescue casualties, etc.

There are many versions of decision support tools – indeed, in the world of management there are hundreds if not thousands of variations of such tools. Those tools most relevant for the FRS IC are the Decision Making Model (DMM), introduced into the UK by London Fire Brigade in the 1990s, The Decision Control Process (DCP), and the JESIP Joint Decision Making model (JDM), which is mainly used for multi-agency incidents and dovetails into the DCP process.

The DMM is gradually being phased out of use in the UKFRS, reluctantly in some cases due to its utility and familiarity/ease of understanding. The JESIP JDM, will be considered in Chapter 15 as part of the interoperability aspects of FRS decision making. The DCP is the current model of decision-making tool favoured by the UKFRS and has been available for use since 2015. It should form the foundation for all operational decisions made on the incident ground, irrespective of the size of the incident, the type of incident and the commander in charge. It is for this reason that the DCP will now be examined in some depth.

The decision control process

The DCP is an evolutionary step above the previous decision making model (DMM) in use for more than 20 years in the UK. It uses the processes associated with the naturalistic decision-making processes discussed above, and much of the DCP is the same as the DMM: a cyclic or iterative process of understanding a situation, planning, implementing and reviewing. What has changed is the recognition that decision making at an incident is not always linear and that steps may be minimised, expanded and reordered in order to maintain progress while minimising risks to firefighters and others. The other key change is an explicit expectation that ensures that controls are placed upon the incident by the IC to ensure that risks are managed proportionately and in line with the perceived potential benefits of a course of action and help reduce the likelihood of the IC falling into a decision trap. The DCP is comprised of four main components:

- An understanding of the situation.
- The development of a plan.
- A review of the plan using the decision control process.
- The implementation of the plan.

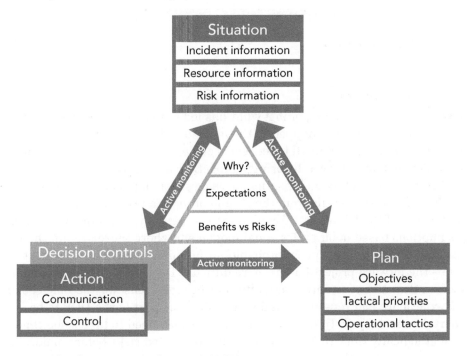

Figure 7.5: The decision control process (DCP)
(Reproduced from NOG)

Component 1: An understanding of the situation

We have already discussed in some detail the processes of information gathering and the development of situational awareness in Chapter 6, which is a key factor in gaining an understanding of an incident. This component of the DCP has a number of familiar elements from the DMM: incident information, resources and risks.

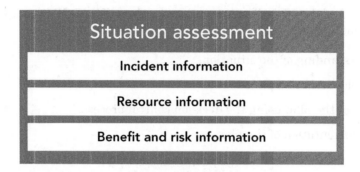

Figure 7.6: DCP Component 1
(Reproduced from NOG)

Incident information

Understanding what is happening, what has led to the current situation, and a projection of what could happen to the incident following various options for intervention is covering all three levels of Endsley's situational awareness model. This will be achieved by bringing together all the available sources of information, including mobilisation messages, personal observation and knowledge, incident details from the IC and others and triangulating the data. It is the role of the IC to sort through often conflicting information (and make presumptions based upon professional judgement where there are information gaps) and come up with a 'most probable' situational assessment. The next stage – the forward projection – is more difficult. Developing likely outcomes based on the current situation and different interventions may become easier with experience and repetition, and will enable this part of the DCP to be undertaken more quickly. Less experienced ICs may need to take some time and methodically work through options and associated risks before making a decision about the actions to be taken.

Resources information

An assessment of resources either in attendance or en route, and those likely to be required to meet the operational requirements of an action plan, should be based upon an accurate forecast of likely developments, plus a contingency reserve to take account of unforeseen developments (such as urgent requirement for reliefs, a tactical reserve and an incident rescue team for more serious incidents). The resources that require consideration include pumps and their crews, specialist appliances, additional officers for command and specialist skills as well as other agencies.

The demand for FRS resources should be based upon the considerations mentioned in chapter 2. In services where crews of four per pumping appliance and where rapid response vehicles are used as part of the 'make up', commanders need to take the potentially lower number of firefighters attending into account and request the appropriate resources. Any request for resources should be based upon a realistic assessment of the incident demands, although erring on the side of caution and having too many resources is always better than having too few. The saying 'you can always send them back' remains true.

Risk information

The risk information required includes:

- the physical risks associated with the incident and the structure/environment which is involved

- the individuals involved (firefighters and members of the public)

- the property that may be affected by the incident.

The sources for this information will include premises data, Hazmat information, occupiers and FRS personnel (see Chapter 6). The additional risks associated with potential interventions must also be considered and factored into the decision-making process. At larger incidents, an ARA may have been completed and this will help inform the IC about both the type and level of risk, but at the early stages of an incident it may be that just verbal information and a DRA is all the information upon which a decision can be based.

Component 2: Developing a plan

The development of a plan is a key phase of any operation – the plan must have an identified goal or objective, tactical priorities and the tactics that will be employed to achieve those goals. Without a clear aim there can be no measure by which to assess the progress of a plan or how it can be determined whether a course of action has been successful. When formulating a plan, the IC must ensure that it is clear, easy to understand, realistic and achievable. At smaller incidents, the development of a plan may take a few minutes; at more serious incidents, more time may be needed. Planning is an iterative process and a formative plan will have been introduced by the initial IC, and that plan, unless it creates an intolerable risk to firefighters, should be continued until a new (and hopefully improved) plan is ready for implementation.

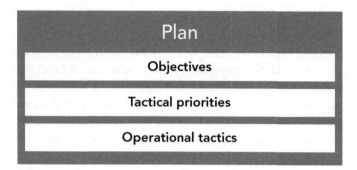

Figure 7.7 : DCP Component 2

Setting objective and goals

For most incidents, the objectives and goals (sometimes referred to as strategic objectives) are relatively straightforward and based around the four FRS aims of protecting life, property, environment and providing humanitarian services, which could be generically described as:

- effect rescues
- contain the incident
- control the incident
- carry out remedial and restoration work.

A clear statement of what the IC considers to be the objectives of the plan, communicated clearly across the incident ground, will help ensure that all personnel are aware of the IC's intentions. This in turn should be used to determine what actions individuals should carry out. If an activity does not help achieve the objective, then it is not required unless there is a very good reason to do it. For example, the use of jets on a building that is likely to be demolished might be a fruitless task but helps reassure the public by giving a positive image of the FRS.

Tactical priorities

The tactical priorities will be determined following the setting of objectives and goals. Let's use an example of a fire in a ground floor shop in a terraced block comprising other ground floor shops and three floors of accommodation and storage above. On arrival there are people waving to the FRS staff from the first and second floors and a well-developed fire is burning in the shop. The objectives are:

- save lives
- control and extinguish the fire
- carry out salvage and remediation.

The tactical priorities will be determined by the level of resources available. If a second appliance is not in attendance the tactical priorities may be limited by the number of crew available on arrival. The tactical priorities will therefore be:

■ carry out rescues from the first floor (nearest the fire and at highest risk)

■ carry out rescues from the second floor

■ attack the fire.

If a second crew arrives at the same time, then the priorities can be changed to take into account the additional resources:

■ Carry out rescues from the first floor.

■ Simultaneously attack the fire and contain the spread (to reduce exposure to fire of the casualty being rescued from above).

■ Carry out rescues from other floors as and when possible (as the risk of exposure to fire will be reduced by the fire attack on the ground floor).

Needless to say that in the examples above there will be many variations and the prioritisation will depend very much on the level of resources in attendance and en route, and the incident dynamics at the time.

Operational tactics

Operational tactics will set out how the tactical objectives will be met. The operational tactics that will be employed to put the tactical priorities into effect should, wherever possible, involve the use of standard procedures developed by services that take into account national guidance and local contexts such as crewing standards, levels of pre-determined attendance and the availability of resources. Operational tactics include the use of BA, ladders, positive pressure ventilation (PPV), ultra high pressure jets (UHPJ), hydraulic rescue gear, etc., to achieve the tactical aims.

The determination of the tactical mode will permit or restrict activities at an incident within the hazard zone, but for most incidents, particularly where rescues are involved, the tactical mode will be offensive (for more on tactical modes, see Chapter 8). If in the unlikely event that standard operating procedures or protocols are not suitable, appropriate or unable to be implemented within a time that supports casualty survival, the use of operational discretion should be considered, providing that additional risks associated with the deployment of new or unusual measures does not raise the risk to firefighters beyond an acceptable level.

If additional resources are required to support the implementation of tactics, then they should be requested as early as possible. Progress of the incident should be monitored by the IC and sector commanders. It is important that the intentions and expectations of the IC are known, particularly by the sector commanders, so that progress can be measured and that variations in the speed of the resolution can be noted or acted upon if necessary. In order to support the monitoring of progress, it can be useful to produce a timeline that sets out milestones for progress. This should be retained with the command support function, providing a visible measure of incident progress.

The incident ground command structure will need to match the tactical requirements to ensure that the incident is safely and effectively managed. This may include the partition of the incident ground through the introduction of sectors and specialist support at various locations.

Component 3: Reviewing the plan using decision controls

Before commencing actions required under the tactical plan, the IC should undergo a systematic review of the overall plan and their intentions. This should not take long as it is intended to be a quick mental check to confirm that activities are proportionate to the situation, that the anticipated actions meet the aims of the objectives and that the risks that firefighters will face are balanced by the potential benefits that may result from a successful intervention. There are three key questions to the decision control component:

- Why are we doing this?
- What is expected to happen next?
- Is the potential benefit likely to be proportionate to the risks of implementation?

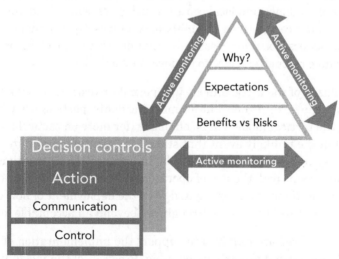

Figure 7.8: DCP Decision controls and the link with component 4 (Action)
(Reproduced from NOG)

Why are we doing this?

This may seem a fatuous question, but it has a serious point to make. How often have firefighters been killed or injured tackling a fire which to all intents and purpose has run its course? Tragically, there have been incidents of firefighters entering derelict buildings, due for demolition, well alight and with no life safety risk, firefighting in dangerous conditions and losing their lives, but to no end. The property is lost and due for demolition, there are no lives at risk (other than their own), and the most pragmatic solution would be to allow it to burn or to control its combustion. The IC should always consider how an action being proposed might contribute to the objectives and goals that have been set. Just asking the question may cause the objectives and goals to be reset. When an objective is being set, the rationale for that objective should be clearly understood and articulated at that time. Where possible, the rationale should be recorded in the decision log so that it may be referred back to if later challenged. Where an objective has been changed as a result of the decision controls being used, then this also needs to be recorded.

What is expected to happen next?

Any actions that are being implemented should have an impact upon an incident. A jet brought to bear on a fire will lead to the expectation that the fire will begin to diminish, and grey or white smoke will be seen. A BA crew sent in to extricate a casualty in a building will be expected to return within a certain period with that casualty.

If the IC can identify key activities and when they should be completed, it will be possible to measure progress against these milestones and identify if things are going more or less to plan, or if things are not working as expected. Ideally, milestones and expectations should be set (and recorded when possible), ideally where they are visible to the command team so that as many eyes as possible are monitoring progress.

Are the risks proportional to the benefits?

While most incidents will be easy enough to assess, remember that the life of a firefighter is too precious to waste on a futile exercise where no possible benefit can be gained. If a firefighter's life is to be risked then the potential benefit of a successful intervention must significantly outweigh that risk. The difficulty is where either the additional risks or potential benefits are uncertain and where a time-critical decision needs to be made. At this stage of an operation, an IC must carefully deliberate and assess the facts as they see them and make a decision based upon those deliberations. The considerations need to include:

- the incident dynamics and potential negative developments (increase in size of fire, failure of part of the building structure)

- the capabilities, skill and experience of crews

- the ability to implement safe systems of work

- the likely survivability of casualties, both with and without intervention

- the likelihood that any intervention will work.

The IC should consider these facts and projections impartially and not allow themselves to be swayed by emotional arguments to undertake dangerous, high risk interventions.

With experience, decision controls can be implemented quickly, almost intuitively, although it is better if a conscious effort is applied. Once the decision controls have been reviewed, the actions to resolve the incident can now be implemented.

Component 4: Implement the plan

Once decisions have been made and a tactical plan determined, the plan must then be implemented. The task is to communicate the plan to those who need to carry out the necessary actions and control the tempo, incident ground activities, and ongoing management of resources for the duration.

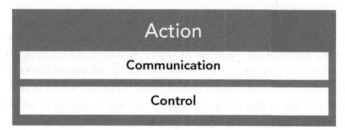

Fig 7.9 Implementing the plan
(Reproduced from NOG)

Communicate

Ideally, this will take place face to face but there are occasions (particularly when plans are being changed or modified) when sector or functional commanders cannot immediately be released to be briefed and tasked by the IC. In these circumstances alternative methods of briefing are required. This may involve the use of incident ground radio networks, the use of a runner with instructions (verbal or written), the use of a phone or a formal briefing by the operations commander – or even the IC if necessary (and support mechanisms in the CSU are available to ensure the continuity of command and control in their absence). The use of a systematic briefing tool such as SMEAC should be used so that the situation, intention and mission (task) are explained thoroughly. Effective communications are discussed in Chapter 9.

Control

Control of the action plan is exercised through the effective use of the command structure, which may include the use of sector commanders, BA supervisors, safety officers and observers all working towards the common objectives set by the IC (see Chapter 8).

Active monitoring

Each stage of the DCP should be actively monitored throughout the duration of the incident. The IC should constantly monitor progress through reporting mechanisms set up either through the adoption of standard operating and communications protocols or through situation reports from all operational sectors, and functional sectors. Importantly, reports are required from safety sector commanders who will be responsible for ensuring that safe systems of work are in place and effective, and that conditions at the incident are developing as expected or deteriorating.

It is also incumbent upon all those on the incident ground that any safety issues are reported as they emerge and that updates on progress are sent regularly. Again, the introduction of a written timeline (in the command unit) with milestones and reporting deadlines for requesting situation reports ('sitreps') and operational updates will help the IC monitor progress, allocate additional resources and request assistance from fire control as they become necessary.

It cannot be emphasised enough that the DCP is iterative and should be continually reviewed, revised and refreshed to ensure that actions on the incident ground remain relevant, safe and effective in achieving the operational objectives and goals.

There are several things that the IC needs to be aware of when considering the decisions being made. These include the potential for tunnel vision, the potential for getting caught in a decision trap and also the impact of incident ground stresses on the decision-making process. We have already considered the issue of tunnel vision in the previous chapter (see page 79), but a decision trap, the result of a failure of thinking things through, can also lead to a dangerous situation or an incident being dealt with inappropriately. Both tunnel vision and decision traps can also be the result of faulty decision-making brought about by the impact of stress on the incident ground.

Decision traps

Decision traps can occur in many walks of life including business, finance and in the emergency services. Many bad decisions can be traced back to failures to collect the right information, options not being evaluated properly and risks and benefits not being weighed and balanced. Sometimes, however, it is the way the human brain works that has the impact on how we make decisions. Researchers studying the way the brain works have identified that we all use unconscious routines, known as heuristics, to help us cope with the complexity that faces us when making decisions. For example, when judging distance we associate clarity of image with the distance an object is from us: the clearer the object the closer we assume it to be. Generally these shorthand systems of thinking work, but sometimes they can cause us problems – in hazy conditions, for example, the brain can trick us into thinking objects are further away. This isn't a problem for some but for a vehicle driver on the motorway it can lead to a catastrophic failure of judgement. Let's have a look at some of these common traps now.

Anchoring

Another phenomenon that we need to be aware of is known as 'anchoring', in which the mind gives disproportionate weighting to the first set of information it receives. The firefighter, given information that a person is reported trapped in a building on fire, may disregard subsequent (and possibly more authoritative) information that this is not the case. There have been incidents where firefighters have been injured in buildings searching for casualties they believe to be trapped, based on the flimsiest of details and rumours from casual observers, such as, 'I saw rough sleepers in that building several months ago!' ICs should always be aware of this phenomenon and the following suggestions may help reduce its likelihood of it occurring:

- Try to review what do you know from different perspectives, especially when additional information becomes available. Question the validity of information you receive, particularly where tactical plans are likely to put firefighters at additional risk.

- Try to think the problems through before consulting others as this will help reduce the likelihood of you becoming anchored by their ideas.

- Remain open-minded and seek the opinions of others to help triangulate information and solutions.

- Try to avoid anchoring your team's opinions and ideas by telling them your opinion immediately. In the same way you might be anchored by others' opinions, they may be similarly influenced and this may lead to 'groupthink'.

The status quo trap

All humans beings are more comfortable accepting a status quo than dealing with change and innovation: the introduction of locomotives, the car, self-contained breathing apparatus and ultra-high pressure extinguishing systems all took time to gain universal acceptance and faced resistance when new. The status quo often represents the safer course and, psychologically, puts us at less risk. In the fire and rescue services, where organisational and individual risk appetites are low, doing something differently may be perceived as 'risky' and unusual, and the lower risk option is often seen as more appealing. Being aware of the status quo trap allows the IC to reduce its influence by taking the following steps:

■ Focus on the objectives and use the DCP (or another model) to identify if procedures are barriers to achieving the overall aims.

■ Always think of alternatives to the status quo option.

■ Avoid exaggerating the disruption caused by changing the status quo option.

■ Ask, 'Would we be implementing this technique/operation/tactical mode if it wasn't the status quo option?'

■ Don't select the status quo because deciding on a number of alternatives is too difficult or time consuming.

There are many examples of this type of thinking process occurring on the incident ground:

■ The defaulting to 2 Breathing Appatatus (BA), I Hose Reel Jet (HRJ) when a main jet/ultra-high pressure jet may be more appropriate.

■ The removal of a roof from a crashed vehicle when the removal of a door and adjustment of the seat would be less traumatic to the casualty and quicker

■ The use of jets on a haystack fire when a controlled burn is more cost effective etc.

The 'If at first you don't succeed, try, try, try again' trap

This phenomenon is also known as the 'reinforcement of failure' trap. How many times have you been to an incident where the fire has failed to go as planned and, instead of stopping and thinking around the problem, the IC has regrouped and then repeated the same operational tactic time and again. Eventually the fire goes out – usually as the result of fuel limited extinction rather than anything the FRS has done. This happens regularly and can be the result of a lack of proper consideration of the alternatives or sometimes as a result of a very human emotion – pride. Changing tactics in the middle of an operation can seem like an admission of getting it wrong the first time and sometimes this admission can be difficult to make. Where things are not going to plan (and we've all had things not go to plan), the right thing is to stop and consider alternatives, generate new ideas and try them out. What can't happen is to persist with a failing tactic at the expense of the incident by failing to reduce risk or save property.

Confirmation bias or confirming evidence trap

There have been many studies about this phenomenon in which we seek out information that supports our existing view or instincts, while avoiding any information that contradicts those views – 'uncomfortable truths'. This is because humans tend to subconsciously decide what we want to do before we understand why we want to do it. We also are more favourably inclined to things we like than things we dislike, and as a result we prefer to listen to people who tell us things we want to hear.

In a rank (or role) based organisation, views that challenge the senior commander's understanding on the incident ground may not be as forthcoming as may be necessary. This can be dangerous as the IC may simply have her/his opinion reflected back to them and reinforced. The self-aware IC should attempt to reduce this bias by:

■ Checking your own opinion and asking if it is reflective of what is or what could be happening.

■ Asking others for their honest opinion. Don't ask leading questions ('This is the way it needs to be done, isn't it?'), and don't restrict your views to your closest colleagues or those whose opinion you respect most – ask a wider circle of opinion if you have time.

■ Ask someone to be a 'devil's advocate' and challenge your opinion, and don't take umbrage if they don't agree with you – that's their job.

The memorable event trap

We all have incidents we attend that are particularly memorable, for good or bad reasons. Sometimes these events have a lasting influence on the way we deal with incidents of a similar type. A firefighter who has had an experience of being disorientated in a large building fire, for example, may be conscious of a potential repeat at a similar incident in the future. A fire that has been fought successfully using a unusual technique (e.g. using an ALP monitor through a glass-fronted furniture warehouse) may influence an IC to try to repeat it at all incidents of that type, irrespective of whether it is appropriate. ICs need to be conscious of the legacy that their 'best jobs' and 'nightmares' have left them, and examine the current incident in a detached manner, not allowing the past to distort their current evaluation.

Other traps and decision-making mistakes

We are now going to have a look at other outcomes that can materialise as a result of the traps above and that can result in other less-than-favourable outcomes.

Relying on intuition: Some types of incident remain fairly common, such as kitchen fires, car fires etc. It is possible that for these 'bread and butter' incidents, the IC can make decisions based upon intuition and experience rather that considering the actual incident at hand. In doing so, the actual strategic objectives and tactical plans may not be factored in to their considerations and as a result there can be a mismatch between the decisions made and the wider objectives for the incident.

Selective information: An incident can be a complex working environment with decisions needing to be made across its length and breadth by commanders at all levels. It is important that commanders keep a sense of perspective by looking at the wider picture, and avoid the temptation to focus on part of the incident alone. We all have our priorities and key issues, but they must not be allowed to detract from the main aim of resolving the whole incident. The example below shows how a focus on a single issue has the potential to disrupt the flow and proceeding of an incident.

Box 7.1: The 'assertive' sector commander

At a major incident involving a large quantity of recycled material, a sector commander responsible for several crews found that the water supply for the jets in his sector were unsatisfactory and required boosting. At the time there were 10 jets in use in the sectors downwind, protecting exposures and reducing the spread of the incident. For a substantial time, the incident command channel was used by the sector commander to justify the request for additional water. Eventually a runner was sent from the command unit to explain the wider implications of the incident and the reason for the water shortage at the time. The sector commander had not only reduced the availability of radio communications for a significant time, his focus on the water issues led to him failing to notice or address the hazards the firefighting teams faced, including significant access problems, falling from height, exposure to smoke, heat and fumes.

Lessons learned

- At large incidents, try to maintain a perspective on the wider picture. Avoid focusing on a single issue to the exclusion of others, which may be of more significance.

- Don't hog the radio channels – they are a safe system of work as important to safety as a breathing apparatus set. Messages should be concise and to the point and not used for conversation.

- If a complex or important message needs to communicated, if the radio precludes its transmission effectively, consider other means such as a runner or a personal attendance at a CSU (having nominated a suitably briefed deputy to provide cover).

Incorrect interpretation of information: Poor situational awareness can lead to an IC (or sector commanders etc.) making decisions that are inappropriate. At a severe fire involving a basement and ground floor of a shop, crews were deployed into an incident that appeared to be under control. Poor communications between the IC and sectors led to improper deployment of BA teams into the building and uncoordinated ventilation, which led to a rapid extension of the fire which trapped and overcame two firefighters.

Decision aversion or 'paralysis by analysis': Failure to make a decision can be a dangerous phenomenon on the incident ground. When, due to the pressure of time, an escalating situation and a high uncertainty about the incident, the availability of several options for action may be available but a decision is delayed due to extensive and time consuming consideration of many options. This may be a result of individual concern on the part of the IC, undue focus on the negative consequences of options (rather than the positive aspects), or an organisational culture where a blame culture exists for failure. An excessive delay in decision making may ultimately increase risks for firefighters, and increase risks to the public, property and the environment.

Failure to follow the DCP: Decision support tools exist to aid the IC by ensuring that objectives, tactical goals and decisions made remain valid and support the resolution of the incident. It is an iterative process that needs to be reviewed regularly to ensure that expectations are being met or that aspects of the plan need to be changed. If the review doesn't occur then the existing plan may become out of date, potentially irrelevant and possibly dangerous as the incident develops and changes. Actively monitoring the incident and reviewing progress against goals will help reduce this risk. Allocating someone from the command team to constantly monitor the plan is essential.

The impact of stress on decision making

Firefighting is a stressful occupation for all involved. At the sharp end, a firefighter on the end of a jet in a building may feel mental discomfort for themselves and their team while an IC may be suffering the stress of having knowingly put others in harm's way without much control over what happens to them. It is important that all firefighters understand the causes and symptoms of stress in both themselves and their colleagues in order to help them cope with the inevitable periods of stress that will occur in their careers and the impact this has on fireground decisions.

Chronic stress is a long-term form of stress and can arise as a result of relationship problems, working in an unsuitable job or workplace, experiencing money trouble or poverty, with little or no prospect of change. It can become ingrained in a person such that they 'live with it', and it can lead to death through suicide, violence, heart attacks and strokes. This is a long term attritional process that can physically wear individuals down.

Acute stress, on the other hand, can be a positive, short-term boost of adrenaline such as is experienced on a parachute jump, skiing downhill or 'tombstoning'. Excessive short term stress, such as the stresses felt by ICs at regular fast moving incidents can, however, lead to a variety of symptoms including:

- emotional distress
- muscular problems including tension headaches, back pain, jaw pain, and muscular tension leading to pulled muscles and ligament problems
- stomach and bowel disorders
- elevated blood pressure, rapid heartbeat, sweaty palms, heart palpitations and migraines.

During an incident, a low-level of stress may give the IC an 'edge', which accentuates awareness and helps clarity of thinking. However, where stress levels are significantly heightened, perhaps due to a commander's 'performance anxiety', there may be an adverse impact upon the IC's ability to command, affecting their behaviours and introducing a risk to the management of the incident. Anyone in the command team or firefighters can be affected by acute stress and all may have an impact on the running of an incident.

We are now going to look some of the factors that can lead to excessive stress being induced.

The incident itself. The incident, the physical conditions and the environment of an incident can have a strong influence on the mood and attitude of those in attendance. Casualties, age, sex and the nature of traumatic injuries can all cause a response in individuals of varying degrees. The time of day, weather conditions and temperature can all help create conditions for acute episodes of stress. The presence of members of the public in distress and any obligation for action may create moral pressure on firefighters, as can an unsuccessful outcome of a rescue operation, leading to a feeling of failing to meet the expectations of a community.

The level of certainty is an important feature of stress at incidents. The more familiar the incident type, the less stressful it may be. Conversely, a new or novel incident, where rapid action is needed or where there are multiple possible solutions, may create uncertainty and a higher level of stress. Where an incident develops in way that is not expected, or where situational awareness is low, uncertainty increases and consequential anxiety may arise. The more complex an incident, potentially with multiple and/or competing goals, the lack of clarity or obviousness of a strategic plan or objectives will add to the confusion and uncertainty for all involved, from the command team to the firefighting teams on the ground.

Where command structures are undeveloped or confused, where there is a lack of control, where freelancing is taking place and information is lacking and situational awareness is low, the acute stress on the IC can be immense, and can impede their ability to command effectively. If command support is missing or new to the role then additional pressure to command the incident is heaped on the IC, adding to an already growing level of stress.

It may be the case that the IC is already preloaded with pressure from the 'day job', i.e. the non-operational environment. With reducing numbers of resources in FRSs, going to an incident doesn't reduce the demand on officers to complete other tasks, and this may have an influence on their levels of stress during incidents. Similarly, an individual's domestic obligations don't stop at the front door: parental responsibilities, caring duties for elderly parents or partners all factor into a firefighter's complex life and can have an impact on their levels of fatigue and stress.

Details of the typical indicators of stress can be found in Appendix B on page 285.

Stress at incidents: impacts

The importance of stress in ICs and others lies in how it can impact upon the management of an incident and not just the effect it can have on the individuals. The distraction caused by the effects of stress can include:

- a failure to maintain a 'grip' on the incident, which can manifest itself practically as a loss of situational awareness

- incapacitated decision-making ability leading to a paralysis in the command unit

- a breakdown in communications, which can lead to misunderstanding the IC's intentions, misdirection of resources or tactical failure to change plans to meet new challenges

- loss of cohesion among the incident command team and incident ground leadership, leading to the emergence of critical gaps in incident knowledge, poor resource management, deployment and lack of effective incident ground accountability

- a compromised or reduced functioning of crew in general, and an increased risk to public safety.

Managing stress on the incident ground

All ICs should be aware of the signs and symptoms of incident ground stress to ensure that it is managed effectively and its impact on operations is minimal. The symptoms of stress are detailed in Appendix B but it is important the ICs are able to control the impact of stress as quickly as possible. This can be achieved by being aware of behaviours that indicate an individual is suffering from the effects of stress, and by paying close attention to the team supporting the commander. Other actions the IC should take to reduce levels of stress include:

- Clear briefings and information exchanges with all staff to ensure they are clear on their mission and tasks at all times. This may involve the use of written briefs or verbal instructions given in a structured way (OTHERS, SMEAC etc.), setting out expectations and intentions, and providing safety briefings from the relevant staff.

- Ensure that the balance of command teams and sectors is appropriate in terms of role and experience, providing additional support for areas or commanders who are likely to experience intense operational demands requiring enhanced levels of supervision. Rotate staff and reliefs at appropriate intervals (based on pressure and level of activities), which may be at periods shorter than normal.

- Create a sense of team spirit, working together to resolve a problem helps ensure that no one is felt to be outside the 'team'. If an incident ground culture of trust and respect is fostered, staff who feel under pressure are more likely to let you know their feelings and concerns. The earlier they make you aware of their feelings the sooner intervention can take place.

■ Personal capabilities and knowledge of the role, service policies and standard operating procedures all aid confidence in operational skills and command ability at all levels, especially when underpinned by regular training at exercises, both in the field and at desk-top level and in simulated incidents.

Individuals should also take a self-awareness check to assess their stress levels during incidents, and there are a number of ways in which these stresses can be managed. Recognising the signs of stress early should encourage you to think about reducing your own levels by one of a number of activities, including finding a relatively quiet area, such as a command support unit's conference/meeting area, and reflecting on the incident and your evaluation of it and your performance. If possible, list what has happened, what is working and what is not. This review of activities and the incident will help clarify key areas of your concerns. It may help to discuss the incident with your team with a view to assessing progress against expectations and to consider the levels of success so far. This will help you objectively assess progress rather than listen to subconscious concerns that may be exaggerating your concerns. If you believe that the stress levels experienced by yourself or others are becoming so serious that they may have a negative impact on the incident or crew safety, you should seek to get immediate reliefs to attend the command support unit.

Summary

You should now understand the following concepts, techniques and practicalities discussed in this chapter:

Taking command at an incident and the first actions to be taken.

The various processes by which decisions are made and how they support each other.

A detailed understanding of the incident management processes including the requirements of the various types of decision-making models including intuitive decision making type such as conditioned processes and recognition primed decision making, rule or procedure-based techniques, analytical and creative methods of determining a solution to operational problems.

A detailed understanding and application of the decision control process including the individual steps in the cycle.

An appreciation of the type of decision trap and other mistakes that can trip up an IC who may be unaware of the decision making processes available.

The impact of stress on operational decision making.

Chapter 8: Organising the incident ground

Introduction

Managing an incident of any type is a complex activity requiring engagement with a range of people, both for the FRS and other organisations, possibly dealing with casualties and occupiers, while still trying to organise rescues, extinguish of fires and save property etc. An effective incident ground organisation and management breaks down this most complex and dynamic of situations into digestible pieces, which allows the incident to be dealt with as efficiently and as safely as possible. The overall incident management system can be divided into two basic components: the system components themselves – the physical arrangement of the incident ground, the communications systems and other assets required to address the needs of the incident – and people and the roles they fulfil to ensure that the system components work together effectively.

This chapter will first consider the development of the physical aspects of incident ground organisation – the concept of spans of control, the sectorisation process, tactical modes, communications networks and the like. We will then set out the key roles supporting the commander on the incident ground and their essential functions – the sector commander, operations commander, hazmat advisor, etc.

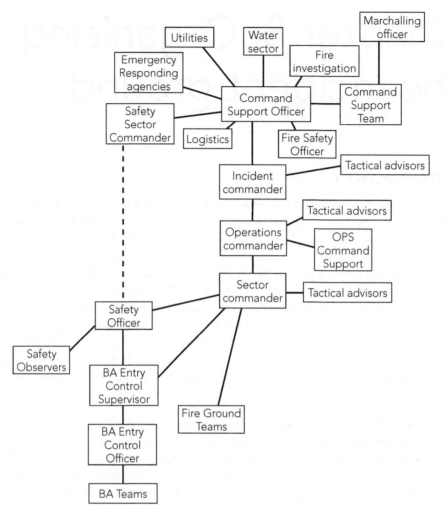

Figure 8.1: incident ground organisation

Incident ground organisation and structure

A good incident command structure for any type of incident is one that enables the IC to achieve the desired results while keeping risks to firefighters and others as low as practicable. Even the largest incident will initially have a response based upon normal requirements for that type of risk. A predetermined attendance (PDA) or Level of Response (LoR) for a typical house may be two pumps, for example; for a hospital fire it may be many more plus special appliances including aerial ladder platforms, forward command units plus a number of command officers. How the incident develops in terms of scale, the relative complexity, its dynamics and a host of other factors will determine the nature of the command structure required. The UK ICS system is a scalable process that can extend to the largest of multi-agency, multi-location incidents and fits into the National UK Concept of Operations for emergencies (see Chapter 15 for an overview). An effective commander will understand the principles and the mechanics of the ICS framework and can then adapt them to meet the needs of any given incident, and they will have a good understanding of the resources available and the capabilities of those resources.

We will now look at how an incident develops from a single pump attendance, building in concepts such as spans of control, sectorisation and tactical modes as the scenario develops.

Spans of control

On arrival at an incident, the initial IC will begin gathering information and confirming the type of incident, carry out a dynamic risk assessment (DRA) and confirm if the resources are sufficient. The IC has a number of direct lines of contact with other parts of the incident – one to the firefighters tackling the incident (via radio or face to face), one to the pump operator (again, face to face or radio), and one to mobilising control (via the pump operator who may also be the initial command support link via main scheme radio). These lines of communication, properly termed 'spans of control', place a demand for attention upon an IC and may reduce their capacity to take in information: information that allows them to maintain 'situational awareness', the vital component of incident management systems. An IC's spans of control will initially be two or three. As an incident grows, the number of spans of control increases.

The concept of spans of control has figures highly within military circles and has done so for centuries. Carl Von Clausewitz recognised the issue as far back as the end of the Napoleonic Wars. He proposed that a commander should not control more than eight distinct 'parts' of an army at any one time, and the smaller the number of spans the better. The downside to this is that with a small number of spans of control, orders may have to go down several layers before they reach the part of the organisation that needs to carry out the activity. This can increase the time before action is actually taken and runs the risk that the message will become distorted. For example, an IC may direct a BA crew to attack a part of a fire in, say, sector 3. This would possibly require the sector commander, the BA supervisor, the BA entry control Officer and the BA crew to be briefed – all of which would take time. An extreme example, but a valid one.

Having a large number of spans of control also creates other problems. There have been numerous examples, both in the military and in the FRS, where commanders have been so overloaded with information that key details of safety critical matters have been missed, misheard or not scrutinised effectively, leading to injury or death. In dynamic situations where the nature, scale and complexity of an incident is changing, the number of spans of control should be as few as is possible, allowing the IC to have the ability to monitor the overall incident and maintain a greater situational awareness that would be possible with higher numbers of spans. For this reason, spans of control should not normally exceed five, and may need to be fewer during particularly fast-moving incidents where critical circumstances are developing.

It is also essential that everyone within the command structure, from firefighter to brigade commander, are trained and know their role and tasks inside out, as the IC is not likely to be able to double check and scrutinise all information that has been the responsibility of others to gather and pass on.

Managing a growing incident

Where an incident is growing or has already reached a size such that the commander has limited or no knowledge of areas of the scene beyond the initial area of activity, it will be necessary to instruct a firefighter to carry out a reconnaissance of the building, sometimes called a 360° assessment, to identify potential rescues, firefighting opportunities as well as risks and hazards in the remainder of the incident area. If operations now become necessary beyond the immediate control of the IC, a second scene of operations will need to be set up. In effect, this partition of operational areas is called sectorisation, which we have just looked at, and has the effect of reducing the IC's spans of control and allows the management and supervision of other areas by sector commanders.

Before 1992, and before the concept of sectors was in widespread use in the UK, control of discrete areas would still occur but in less formalised terms. A 38-pump hotel fire in Maida Vale in London in December 1974 necessitated multiple rescues from front and back of the premises. Sectorisation took place, albeit with a seemingly casual nomination of an officer – 'I met the Assistant Divisional officer and asked him to take charge of firefighting operations and searching at the rear of the premises' (Honeycombe, 1984). This was less formal than would be expected today, but the effect was the same. Spans of control were reduced and this allowed the IC the time to consider wider aspects of the incident and gave sector commanders the ability to closely supervise firefighters within their area of responsibility.

Sectorisation

A sector can be considered to be either a geographical location or a function on the incident ground. It is a subordinate part of the organisation and is delegated responsibilities and authority from the IC. Apart from the maintenance of manageable spans of control, sectorising an incident will have a number of advantages when it comes to its management:

- A geographical division allows a high level of control where there are many activities occurring simultaneously within a sector.

- It improves the flow of communications across the whole of the incident by filtering out sector-specific communications and allowing the IC to be made aware of key issues only.

- It helps ensure the safety of firefighters by having a more concentrated focus and supervision in a relatively small sector.

There are a number of reasons why the creation of sectors may be required:

- To improve delegation of responsibilities, which will reduce the spans of control, particularly when there is an escalating incident that threatens to overwhelm the IC's capacity to communicate effectively with subordinate activities and locations.

- It will be self-evident from the start of some incidents that they will become major events requiring the mass deployment of resources and necessitating an effective incident command structure. Where this is the case, the initial IC should recognise that creating an incident ground command structure is the priority rather than taking immediate, short-term actions (other than immediate rescues) which may impact on subsequent developments, particularly with regard to the overall outcomes of the incident.

- Sometimes it may be necessary to introduce an additional level of command, even though the geographical requirement for a sector may not be required. For example, if a single sector is in operation but large numbers of BA are in use, it may be expedient to position a breathing apparatus operations sector (with associated command and support functions) within that sector.

- Where crews are working in relatively remote locations, away from the main area of operations, sector command will allow effective control and supervision to be implemented and a higher level of relative safety maintained.

Each sector is managed by a sector commander with delegated authority from the IC to use resources within that sector to meet the objectives in line with the incident plan. Operational sectors generally involve the management of the geographical location or portion of a building with the intention of carrying out operational tasks focused on the extinction of fire, rescues etc. Operational sector commanders normally report to the IC, or the operations commander in the case of larger incidents. Functional sectors, which may include a water sector, a decontamination sector etc, will report directly to command support, thus maintaining the IC's spans of control to a manageable level. Not every incident requires sectorisation but very often incidents beyond three or four pumps (i.e. 16 or more firefighters) will require the introduction of sectors in order to manage the incident effectively.

It is important that sectors are not created 'just because we can', but that they add value to the command process. Complicating an incident management structure unnecessarily uses up vital resources, gives rise to the potential for miscommunications and an over-managed incident.

A single scene of operations or the use of two sector (front and back or sectors 1 and 3 – see below) is normally sufficient to manage a serious but straightforward two, three or four pump house fire (See Figures 8.3 and 8.4).

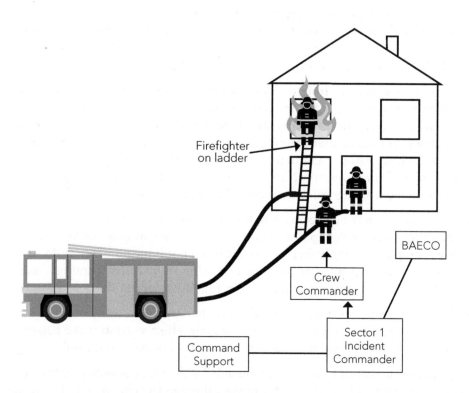

Figure 8.3: House fire – single scene of operations

As an incident grows, the command structure increases in size and complexity. The UK FRS has a adopted a standard method of numbering sectors. Taking the location of the first arriving resource to an incident as 'Sector 1', subsequent sectors are numbered in a clockwise sequence as in Figure 8.4 below. Using a standardised system ensures that reinforcing resources, no matter which service they belong to, can be directed immediately to the appropriate location following a briefing.

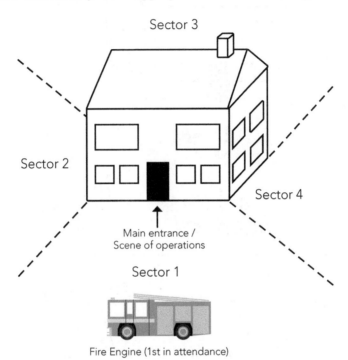

Figure 8.4: UK standard methodology for sectorisation

With the creation of a second sector (sector 3, above) and the nomination of a 'sector 3 commander', a question arises as to what now happens in sector 1: does the IC create a 'sector 1 commander' and also retain the role of IC or do they take both roles – managing sector 1 and the overall incident. The answer depends, usually, on the nature of the incident. Using spans of control as a tool, by delegating sector control to others, the IC's spans of control are three – sector 1, sector 3 and command support. If the IC remains IC and sector 1 commander, the number of spans can be the same or greater – sector 3, command support, BA ECO, plus any additional resources or agencies arriving at the incident. So, for a two, three or four pump incident, the IC can determine the structure depending on the needs of the incident (see figures 8.5 and 8.6 below).

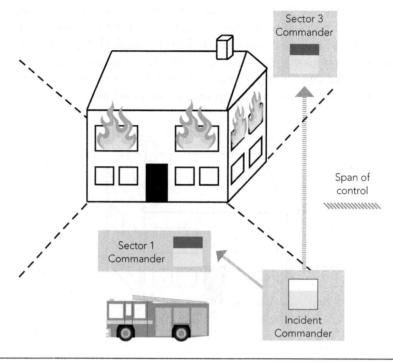

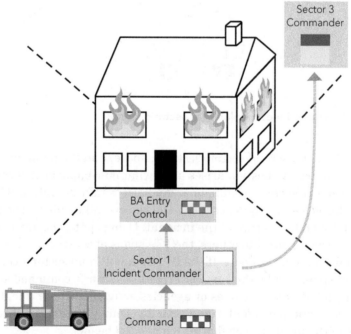

Figures 8.5 and 8.6: Significant house fire – alternative command structures

As an incident starts growing and the number of discrete areas of activities increase there will be a need to introduce measures to maintain spans of control to a manageable number. This may involve the introduction of additional sector commanders, the command support function that will oversee support sector activities through functional support sector commanders, or the use of an operations commander. It is important to remember that the spans of control of these sector commanders should be managed and, depending upon the activity level, kept to a maximum of five, under normal conditions.

The two examples below show an outline of how incidents with various levels of resources (6 and 16-20 pump incidents) can be organised to ensure that the principle of managing spans of control can be arranged.

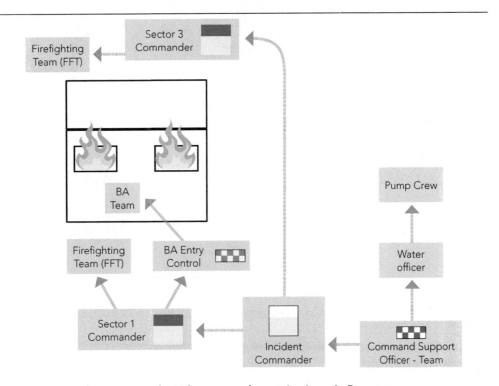

Figure 8.7: Sectorisation and incident ground organisation: six Pumps

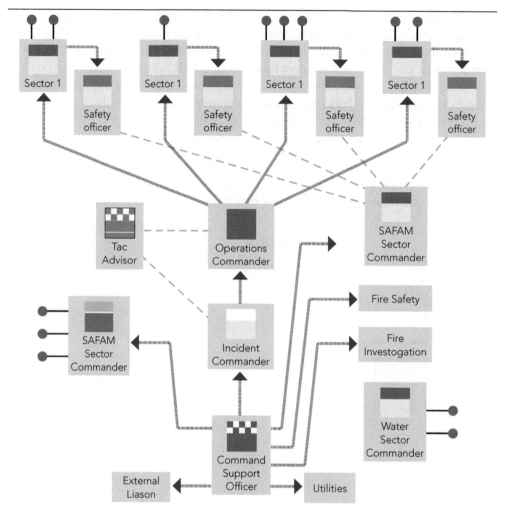

Figure 8.10: Sectorisation and incident ground organisation: 16-20 pumps (large incidents and 'major incidents')

Special considerations

There are several situations that can arise that do not always fall into the standard pattern of incident growth and have the potential to cause difficulties in terms of incident ground management. These circumstances can include the development of a fire beyond the footprint of the first fire but clearly caused by spread (embers or radiated heat) from the initial fire, or when an event (such as a collapse of a wall) causes the separation of a single sector into two mutually inaccessible parts. There are a number of solutions to both.

A secondary fire

The most straightforward solution to this problem is to create a separate incident, complete with its own attendance and command structure. There are a number of complicating factors that the IC should take into account:

■ The time that the 'new' incident PDA will take to arrive, bearing in mind the likelihood that service resources may be depleted as a result of the first incident and attendance times will be extended.

■ The speed of growth of the fire.

■ The potential confusion for resources arriving at a location with two fires.

■ Doubling of incident command radio channels and creation of new 'IC' call signs etc.

These practical considerations aside, creating a separate incident can help minimise the spans of control for the IC.

Box 8.1: A secondary fire at a major mill fire

During a night-time fire at a mill complex, 200m x 200m in scale, large flaming embers were noticed landing on adjacent properties, 100m down wind and starting small fires on the flat roofs of terraced, multi-storey buildings of modern design. Because the incident was within the proximity of the original incident and the delay in the arrival of reinforcing appliances caused by the demands of the incident (30 pumps, four aerial ladder platforms (ALP) plus other special appliances and vehicles) necessitated the allocation of two pumps to deal the incipient fires, it formed part of the overall incident structure and was designated 'Sector 5'.

An alternative solution could be to add an additional sector to the original incident and manage it through existing command structures. This will have the advantage of a more rapid response to the second incident using some resources already available at the first – at least carrying out a 'holding action' while additional support is being mobilised. Again, spans of control need to be tightly managed and the incident ground organisation needs to be carefully considered and implemented to avoid potential confusion.

Box 8.2: Two large fires within the same industrial estate

A fire in a large warehouse containing waste rags for processing caught fire. Within 20 minutes, a large warehouse opposite the first unit caught fire. Due to the potential confusion of having two large incidents within 30 metres of each other, the incident commander sectorised the incident as shown in Figure 8.12. The advantages were that the IC spans of control were kept low, each 'incident' had an operations commander dedicated to it, resources were pooled and allocated by the IC, and only one command team was required

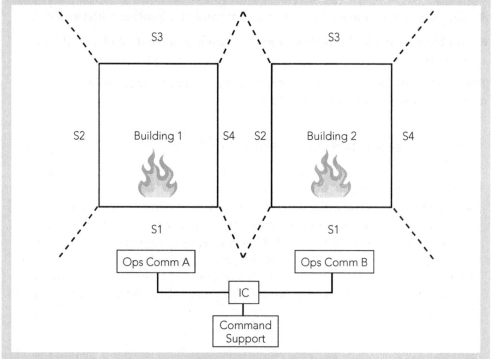

Figure 8.12: Sectorisation – two large fires in very close proximity (one possible solution)

Collapse of a wall/access problems/extensive boundary incidents

Where sectors have become extended so that adequate control cannot be exercised by a single sector commander, it is worth considering dividing the sector to create additional sectors. There are a number of ways in which this can be achieved but the designation of a sector must be considered before implementation. The arbitrary designation of a 'sector 5' which falls between pre-existing sectors 3 and 4 can create uncertainty as firefighters have become familiar with the clockwise designation methodology and sector 5 may be logically considered to fall between sector 4 and sector 1. It may be easier to form sectors from an existing sector so that a collapse of a wall in sector 3, for example, would allow the creation of sector 3A and 3B, keeping the clockwise nature of sectorisation.

Box 8.3: Obstructions/access difficulties on a sector boundary

A large fire involving large quantities of waste abutted onto a building which made communication and access across the sector impossible. In order to avoid re-numbering the sector or introducing sector numbers out of sequence, a solution was used as detailed below.

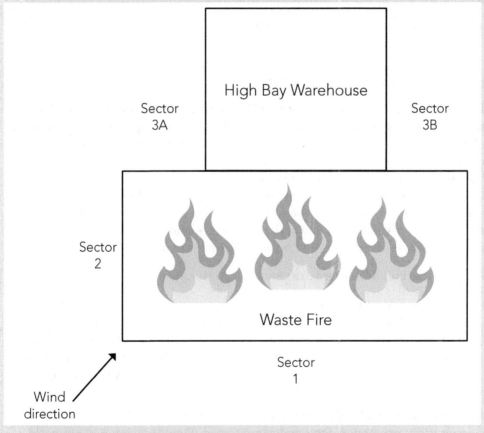

Figure 8.13: Obstructions across a sector perimeter (a possible solution)

Specific types of sectorisation

While the approaches to sectorisation detailed above can be applied to the majority of incidents, there are a number of specific types that require special consideration due to the type of premises or the nature of events. These are incidents involving:

- fires in high rise buildings
- basement fires
- hazardous materials incidents
- road traffic collisions
- wide area incidents – flooding and wildfires

These each need to be considered separately.

Fires in high rise buildings

High rise buildings, above six floors or 20 metres, create a number of problems when responding to an outbreak of fire, particularly on upper floors where the use of aerial appliances is restricted due to the height or where access for the vehicles at the ground floor is difficult due to parked vehicles, road width etc. Access to the building for firefighters can also be restricted, which means that sectorising the building as one might for a low rise building is generally not possible, and a specific method of organising the incident is required. There is also the possibility of high levels of congestion and activity at bridgeheads and entry control points due to restricted floor space and access and egress points.

For high rise buildings there are usually a minimum of five areas of activity:

- **The external approaches** to the building, encompassing the hazard areas immediately surrounding the building and its approaches, and the staging area for appliances and equipment. This is the equivalent to the outer cordon and should be controlled by FRS or police officers. If winds speeds are high, there is the risk that planeing glass may drift some distance and cordons should be in place to reduce the risk of injury.

- **The bridgehead** is the location from which all firefighting and search operations are undertaken and commanded. The bridgehead is normally located two clear (smoke-free) floors below the fire floor. Where fire conditions are such that smoke penetration exceeds two clear floors, then a location more than three or more floors below the fire floor should be considered.

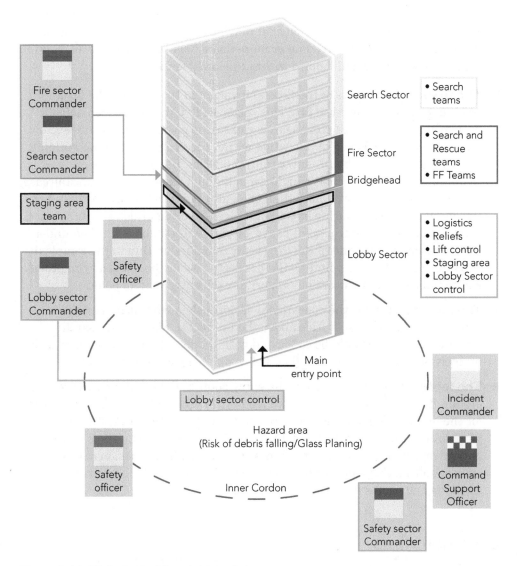

Figure 8.14: High rise incident schematic layout

Where extended operations are taking place at a high level and equipment and firefighting resources are extensive, it may be necessary to site equipment and firefighters in a staging area one or more floors below the bridgehead, possibly using corridor space and lobbies within the lower levels.

■ **The fire sector** consists of the floor(s) involved in the fire and the floor level immediately above the fire. This sector will be the one responsible for firefighting and for conducting search and rescue operations. The fire sector commander will be based at a location within the bridgehead of the building and will be supported by BA entry control officers, BA supervisors and communications staff. When a high-rise building has more than one staircase, it may be possible to have more than one entry point to the fire sector. Where this is the case, the separate entry points may be designated as 'entry control point alpha', 'entry control point bravo' etc, but will still be controlled under the operational command of the fire sector commander. Due to the complexity of using multiple entry points to the fire sector, it may be necessary to provide additional command officers in support at each entry control point.

■ **The search sector** is the operational sector tasked with search and rescue, ventilation and other activities above the fire sector. Ideally, the search sector commander should be co-located with the fire sector commander at the bridgehead, but may be on the floor below due to space limitations.

■ **The lobby sector** extends from the entry point to the building, through the ground floor lobby and to the bridgehead sector. The lobby sector is responsible for the marshalling of firefighters, resources and equipment that are required for operations at the bridgehead, fire and search sectors, as well as ventilation and salvage and co-ordinating the internal onward transmission of evacuees and casualties. The lobby sector commander is responsible for controlling access to and egress from the building.

Basement fires

Basement fires are not only some of the most difficult to tackle operationally, but they are inherently challenging from a command and control perspective. In many ways they are similar to high-rise building fires in that there will be search, fire and lobby sectors, but here these sectors are inverted. While the principles of sectorisation set out below can be applied in all basement properties, they are particularly relevant where there are multiple basements in commercial and industrial premises. The use of sub basements in large residential properties in expensive city centres is becoming increasingly common and may be up to four subsurface levels served by a single staircase. For a single basement premises, the fire will be dealt with from the ground floor level where the fire sector will be located. The search sector, if required, will extend from the ground floor to the top floor of the building, i.e. the floors above the fire. Where there are multiple basements and the fire is not on the lowest basement, then an additional search sector needs to be instigated to carry out a search of the floors below the fire. For all fires involving basements, it is worth considering the deployment of additional safety officers to monitor the safety of crews working in arduous conditions, particularly when using BA.

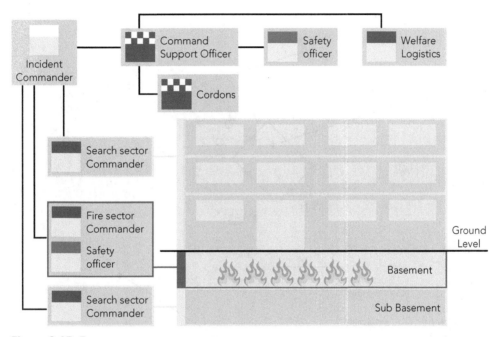

Figure 8.15: Basements

Sectorisation schematics for other types of incident

Hazardous materials incidents

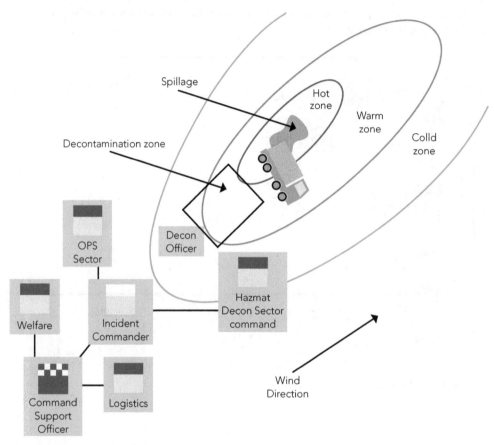

Figure 8.16: Hazmat incidents

Road traffic collisions

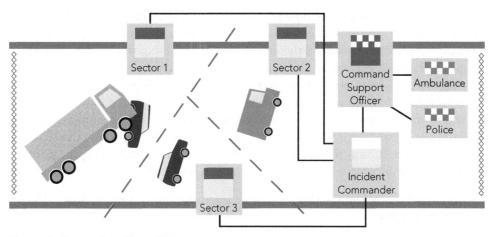

Figure 8.17: Road traffic collisions

Railway incidents

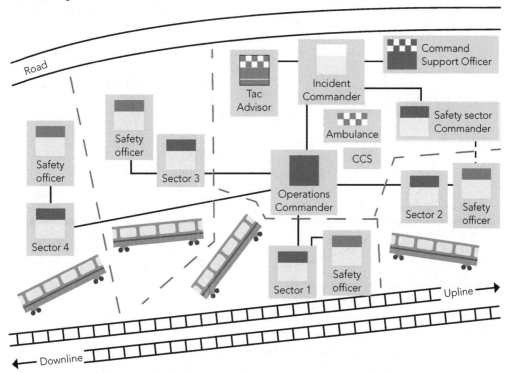

Figure 8.18: Railway incidents

Hazard area and cordons

The hazard area encompasses those parts of the incident ground where firefighters are placed at a significant risk of harm. This could be an area, including the interior of the premises, where there is a risk of building collapse, exposure to fumes from chemicals or smoke, explosions, the planeing of glass and debris from tall buildings. Defining a hazard area, particularly in the early stages of an incident, can be difficult and it is often the case that an extensive hazard area is designated initially and reduced in size as more information becomes available and risk assessments are more reliable. Responsibility for defining the hazard area remains with the IC at all times. In this, the IC may be supported by the safety sector commander, the safety officer and the safety team (observers) in determining the extent of the hazard area.

The **hazard area** is bounded by the **inner cordon** and controls access to the scene of operations.

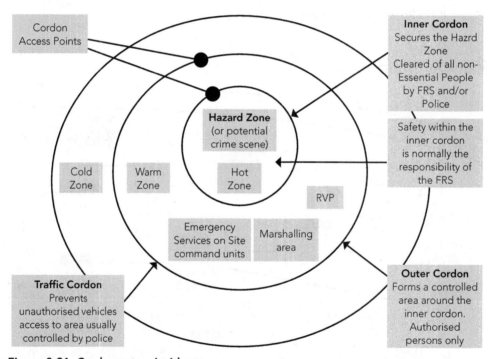

Figure 8.21: Cordons at an incident

The **outer cordon** controls access to the area(s) being used by the emergency services and other agencies. Where practical, the police, in consultation with other emergency services and specialists, will establish and maintain cordons at appropriate distances. Cordons are established to facilitate the work of the emergency services and other responding agencies in the saving of life, the protection of the public and property, and the care of survivors. Marshalling areas, command support and logistics functions/equipment dumps will be located within the outer cordon area.

Where terrorist action is suspected to be the cause of an emergency, the police will take additional measures to protect the scene (which will be treated as the scene of a crime) and will assume overall control of the incident. These measures may include establishing cordons to restrict access to, and require evacuation from, the scene, and carrying out searches for secondary devices. If it is a terrorist incident, the police will ensure that health and safety issues are considered and this will be informed by an assessment of the specific risks associated with terrorist incidents.

NB. In some areas, there are agreements between the FRS and the police for controlling entry to cordons. Where this is the case, fire and rescue personnel are trained and equipped to manage gateways into the inner cordon and will liaise with the police to establish who should be granted access and keep a record of people entering and exiting.

Tactical modes

A tactical mode is the formal declaration of the type of operations that will take place within a sector or incident following an assessment of the risks and the development of an operational plan. The formal declaration of mode to Fire Control and its transmission is deemed by the HSE to be sufficient proof for legal purposes that a risk assessment has been undertaken and the outcome recorded. It also confirms that a safe system of work has been identified and implemented.

The determination of tactical mode is a responsibility of the IC and takes into account the balance of risks versus expected benefits for any given course of action. It is this determination that will determine whether firefighters undertake actions within the hazard zone or outside it.

Essentially, an offensive mode in a sector should be declared when firefighters are working in the hazard area and a defensive mode when they are working outside the hazard area. Each IC at whatever level in the organisation must make this determination, ideally using the decision control process or similar (see Chapter 7) as soon as possible, and ensure it is communicated to the whole incident ground and recorded, either through radio message to mobilising control or by written means.

'Offensive' and 'defensive' are the only two tactical modes available. They can be applied to the whole of the incident or individually to each sector. If all operational sectors are in offensive mode the term 'offensive mode' will be assigned to the whole incident. Similarly, where all sectors are defensive, the incident will be declared 'defensive mode'.

In addition to the declaration of tactical mode, it is important that additional information about the justification for the mode is given where possible. Examples of messages include:

- 'Crews in offensive mode: persons reported or saveable life.'
- 'Crews in offensive mode: saveable property.'
- 'Crews in offensive mode: environmental protection.'
- 'Crews in defensive mode: awaiting isolation of power cables.'

Offensive mode

A declaration of offensive mode means that firefighters are working within the hazard area as defined by the IC. This means they are working in the area of heightened risk. At a majority of incidents, it is likely that most of the time firefighters will be working in the hazard area and that the incident will be declared as being in 'offensive mode'. This will normally be the case at residential and other structural fires as well as road traffic collisions where rescues and firefighting is likely to be taking place. It is important to note that even where the firefighting strategy may be defensive – that is, the building on fire has been viewed as non-saveable, but surrounding properties and exposures need to be protected with firefighters working within a hazard area – the tactical mode is still offensive because firefighters are within the hazard area while playing jets on the exposed property. In determining the 'Hazard Area', the IC may need to use the advice of specialists (Hazmat, Urban Search and Rescue or National InterService Liaison Officer advisers, for example), or safety officers who will have a greater capacity to assess and evaluate risk levels and threats to safety.

Defensive mode

When crews are not working within the designated hazard area, the sector is in defensive mode, due to the risks of operating within the area being unacceptable, even when additional control measures have been implemented. For example, the presence of compressed gas cylinders, evidence of instability of the structure (cracks in walls, bowing out of gable ends, penetration of structure by expanding steel beams), and excessive smoke blowing into the hazard area are all situations that could necessitate withdrawal of firefighters.

Examples of a defensive mode include waiting for the arrival of specialist advice and assistance such as the isolation of utilities or standing by before committing to an area where terrorist acts may have occurred or are anticipated. It is still possible, however, to undertake 'offensive operational tactics' while in defensive mode, as firefighting action can take place from outside of the hazard area. An example of this is where an aerial ladder platform (ALP) is being used as a water tower to extinguish a building fire with the vehicle being placed a significant distance from the fire. When monitors are in use, the mode will only be defensive as long as no firefighters are in the hazard area. If it becomes necessary to reposition the monitors, then the incident will be in offensive mode while firefighters are in the hazard area.

No overall mode

Where offensive operations are being undertaken in one sector and defensive operations are being undertaken in another, i.e. incidents where both tactical modes are being used simultaneously, these are said to be in 'no overall mode'. Messages to fire control and across the incident ground should confirm that the incident is in 'NO OVERALL MODE' followed by a list of each sector and confirmation of the mode it is in. For example, 'Incident in no overall mode: Sector 1 – defensive mode; Sector 3 – defensive mode; Sector 4 – offensive mode'.

Changing tactical mode

The tactical mode in a sector may be changed for a number of reasons. Where the risk to firefighters has been reduced (for example when defensive firefighting operations have reduced the intensity of a fire sufficiently to allow firefighters to enter into the hazard area) the mode may be changed from defensive to offensive. If the reverse is true and conditions have deteriorated such that risks in the hazard area have become too great, then the intention of the IC will be to change the mode in that sector (or whole incident) to defensive. Withdrawing firefighters from the hazard area can be achieved in one of two ways: a tactical withdrawal or an emergency evacuation. Irrespective of which option is chosen, the *tactical mode will not change to defensive mode until those firefighters have withdrawn from the hazard area to an area of relative safety*.

Whenever the tactical mode is changed, the commander should ensure that all parts of the incident ground and fire control are notified. Furthermore, confirmation of tactical mode should be notified both to fire control and the incident ground on a regular basis. Each service will determine its own frequency of confirmation of tactical mode but should be not more than every 30 minutes.

Tactical withdrawal

A tactical withdrawal should be ordered as soon as the risks within a hazard area have been deemed too great to justify firefighters remaining within that area. When it has been decided that a tactical withdrawal is required, the relevant sector commander should be ordered to begin the withdrawal and to notify the IC (or the operations commander) when the withdrawal is complete and all firefighters have left the hazard area. This will then enable the message indicating the change of tactical mode to be sent. When the decision to undertake a tactical withdrawal is made and the order given, a message indicating **'tactical withdrawal in progress'** should be sent to fire control to record the changed outcome of the dynamic risk assessment at the incident. On completion of the withdrawal, the change of tactical mode should be notified to fire control.

Emergency evacuation

In certain circumstances, for example, a rapid increase in the intensity of a fire, or new information regarding risks within a premises or vehicle etc. becoming available, it may be necessary to carry out an urgent removal of firefighters from areas where the risk has become too great. This 'emergency evacuation' should be notified to the area using fire ground radio channels and using the words, **'All sectors evacuate the incident ground now: evacuate, evacuate, evacuate'.** Fire control must be notified of the emergency evacuation through an informative message using the phrase **'emergency evacuation in progress'**, which will record the change in the dynamic risk assessment. Any emergency evacuation message should be accompanied by the use of repeated blasts on evacuation whistles, usually carried by all operational firefighters. As soon as practically possible, a role call should be undertaken of all crews, initially by the sector commander in their sector where sector accountability boards are available. Where this proves problematic, then crews should be directed to the incident command unit or emergency evacuation point where there should be a record of all firefighters attached to the incident.

Emergency evacuation points

At incidents where the emergency evacuation of personnel has been identified as a possibility in the early stages, part of the risk control process may be to have designated an emergency evacuation point, which is remote from the immediate hazard area and allows a rapid role call to be taken to identify missing personnel. The emergency evacuation point should form part of the initial safety brief for firefighters so that everybody knows where to report in the event of an emergency evacuation signal being given.

The FRS is responsible for managing the health, safety and welfare of those who may be affected by operations within the risk and hazard area. As part of these responsibilities they should be considering the impact of the incident and FRS actions on the safety of other agencies' staff. Accordingly, the staff should be properly briefed and notified of emergency evacuation points and evacuation procedures on arrival at the scene of operations.

Pressing the 'reset' button

There may be times when the IC arrives at the incident and that 'gut feeling' or other indicators suggest that all is not well on the ground. It may be that information from the many sources do not triangulate and there is a disconnect between expectations and reality. By gaining a sense of the incident – seeking 'ground truth' – the IC may consider that operations have moved beyond the point where the balance of risk and potential benefits has become distorted and that firefighters are in danger. In this instance, the options are limited and may require an immediate withdrawal, followed by a period of re-evaluation and carrying out an assessment of the current plan, its suitability and the possibility of any modifications (or abandoning the current plan in its entirety). In effect, this means starting the planning process again. While this may be a radical and sometimes unpopular decision, the IC must be in a position where the plan is appropriate for the incident and any discomfort they (or their firefighters) have regarding a drastic change of tactics (or strategy) is secondary to the safety of crews.

A more difficult situation arises when the oncoming IC only believes that things aren't right but there is limited evidence on which to base that impression. At this point a careful analysis of the incident is needed, using a decision support tool (the DCP or DMM, see Chapter 7) to gather information and reformulate a plan of attack. Where there is any suspicion that the risk to firefighters is excessive then a tactical withdrawal should be considered. Although this may be one of the most difficult decisions an IC may have to make, the failure to do so may result in far worse outcomes.

Probably the hardest decision an incident commander will have to make is that which is made in the knowledge that a casualty remains within a building where the fire is not survivable. Despite the fact that there is no benefit to continuing rescue operations and the risks are intolerable for firefighters themselves, there is often the temptation to continue the rescue attempt. While this most human of instincts is praiseworthy, the IC must remember that where the balance of risk and benefit is so imbalanced, they must halt rescue operations until the balance is recovered.

Box 8.4: A controversial retreat

At a serious fire in the roof space of a three storey Victorian school, two BA teams were attacking the fire and believed they were at the point of its extinction. The initial incident commander carried out his 360° and believed the incident was under control based on the information received from the BA teams. A more senior officer attended the incident and considered the firefighters in the roof space were at risk of becoming trapped should the fire enter one of the many void spaces within the roof space. After a short discussion, the senior officer took command of the incident and ordered the withdrawal of the BA teams and declared the incident tactical mode as being 'defensive'. The BA teams, having believed the fire was on the point of extinction, were frustrated with the withdrawal decision. A short time later, the fire, which had entered ventilation ducts in the attic, did indeed break out in several locations on the third floor and the roof space, causing local collapses and a rapid spread of the fire. The building was severely damaged and eventually rebuilt.

Lessons learned

- Pre-incident knowledge of the building structure would have helped the initial IC identify the risk of rapid spread.
- Additional safety observers and BA safety teams would have helped the IC to monitor fire conditions and assess fire development.
- Crews within a building cannot always have a comprehensive understanding of fire conditions and the IC needs to ensure that a wider perspective of the incident is used rather than relying on single sources of information.

Fire and rescue service personnel roles

It is important for any IC to be familiar with the different roles that occupy incident ground and the functions they carry out in support of the resolution strategy. These roles can vary with each incident and the specific tasks and objectives the IC wishes to be achieved.

There are several categories of role on fire ground:

- Command based roles.
- Support sector or functional roles.
- Tactical advisers.
- Non-FRS support roles.

When communicating with the IC, operational sector, functional sector, or any role on the incident ground, it is important to use the correct terminology which is a role designation and not use individual names, appliance call signs or informal designations. When calling on the radio for the sector 1 commander, for example, they should be referred to as 'sector 1 commander'. Similarly, 'operations commander', 'water sector commander' and 'command support' should all be used as call signs.

We are now going to look at some of these roles in detail.

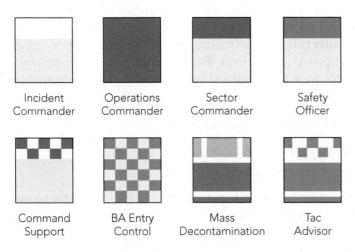

Figure 8.25: Fireground identification tabards for key command roles

Command roles

Operational sector commander ('Sector commander')

Sector commanders managing an operational sector will be responsible for commanding resources within a geographical location and directing activities such as deploying crews and managing operational tactics in order to meet the operational strategy and tactical plans set out by the IC. As part of this plan, they will be responsible for briefing crews, ensuring safety, setting out tactical priorities and ensuring that the resources in this sector are sufficient to carry out any tasks allocated to them. The organisational position (or role) of the sector commander should be such that he/she is competent and with such authority that insures that instructions and orders cannot be challenged by those being so directed. It is the responsibility of the IC to ensure that sector commanders appointed are of appropriate organisational position and authority, and are competent in the role being assigned.

It is important that resources within the sector are not so great that the sector commander is overwhelmed and their spans of control excessive. It may be the case that a sector command support function is created to manage the sector commander's communications with the IC and/or command support unit, update sector accountability boards or any other activity that helps ensure that the sector commander does not exceed their limit of spans.

Sector commanders should, where possible, be located at a sector command point that is easily visible, recognised and with good communications with the command support unit. Fire crews allocated to that sector should report directly to the sector commander for safety, tactical and communications briefings, as well as sector orientation before undertaking any tasks. The sector commander (or sector command support) should be in direct contact with the crew or watch commander responsible for a team within their sector. The specific responsibilities of the sector commander are to:

- brief crews and direct sector activities as required

- monitor the work of crews within the sector

- monitor the personal safety and welfare of individuals

- ensure resources requested are sufficient to carry out allocated tasks and to ensure that reliefs are in position and ready for deployment when needed.

- ensure that the transition arrangements for relief take place smoothly – ideally, one crew at a time rather than whole sector being relieved simultaneously

- maintain communications with the IC and adjacent sectors to ensure that operational tactics are fully aligned do not compromise crew safety

- regularly update the IC with progress against plans with a situation report (SitRep)

- ensure that all crews are debriefed before dismissal from the sector.

Where appropriate, ICs should confer with sector commanders when making a decision to change the tactical mode. **It is more usual for the suggestion to change tactical mode to come from a sector commander**. Sector commanders may change the tactical mode used within their sector with the formal an explicit approval of the IC (or the operations commander where one has been appointed).

There may be occasions when they need to act first in the interests of safety and then inform the IC/operations commander of their decision. Where it is necessary to remove firefighters from a hazardous area using a tactical withdrawal, approval for such a move is not required immediately but notification to the IC needs to be made as soon as practicable. The tactical mode should then be determined through discussion with the operations commander and notified to fire control and the incident ground.

If a sector commander wishes to commit personnel into the hazard area, for example make a change to offensive mode when the prevailing mode is defensive, they should first seek permission from the incident or operations commander. They should not make any change until they have received permission.

Where a rapid change in circumstances occurs, the sector commander should revise the risk assessment. Upon taking command when a sector has been created for the first time, it is important that a dynamic risk assessment (DRA) is carried out for that sector, supplemented as soon as practicable by a sector specific analytical risk assessment (ARA) completed by a suitably competent person. The sector commander is responsible for ensuring the regular review of the ARA and the transmission of its key findings to the incident safety sector commander for compilation and to inform any actions that may be necessary.

Sector commanders may be supported in their sectors by breathing apparatus entry control point officers, BA supervisors and safety officers/safety observers. These roles are described below.

It may be the case that, on occasion, a command team briefing is required which may necessitate the sector commander (and other sector and functional officers) attending a single location such as the command support unit. In order to ensure continuity of sector command, it is necessary to appoint a replacement with equal command skills and (ideally) authority, on a temporary basis until the conclusion of the briefing. When this occurs, it is important that all personnel within the sector are made aware of the change and that the command support unit is advised and the incident command structure record is likewise amended.

Operations commander

The role of the operations commander is somewhat discretionary up to a point: as long as it is possible to maintain spans of control at five or less, command through the IC could be relatively easily controlled. Where the number of sectors threatens to exceed four, the spans of control are at a maximum (five: four sectors and command support) and where there may be additional demands for attention, for example multi-agency meetings, briefing political leaders, principal officers etc. the role of the operations commander may be activated to undertake the management of operational activities within sectors.

The role of the operations commander is to work as part of the command team, which also consists of the IC and command support officer (and in some services the safety sector commander), and:

- co-ordinate the operational activities of each sector

- co-ordinate resources, reliefs and welfare

- ensure all risk assessments are completed, reviewed regularly and updated, and forwarded to the command unit for collation, recording and upward transmission

- co-ordinating communications between IC and sectors

- determining the appropriateness of the tactical mode and approving any changes of mode.

Operations commanders should not concern themselves with activities outside the operational activities within sectors. 'Mission creep' should be avoided and contact with support and functional sectors (media, fire investigation, fire safety etc) should be limited to essential issues that directly impact on the operational activities within sectors.

At exceptionally large incidents such as wide area flooding, wildfires, multi-location terrorist events and major fires, it may be necessary to appoint more than one operations commander. Where this does occur, it is important that the designation and call sign of the operations commanders are distinct and easy to recognise. Radio call signs may be predetermined ('operations commander **alpha**', 'operations commander **bravo**') or organised by the part of the building involved ('operations commander EAST/ WEST/BUILDING One/Two etc.) (see Figure 8.12). At these more complicated incidents, is important that all personnel on the incident ground are briefed on the command structure and under whose command they fall.

Breathing apparatus entry control operative (BAECO)

At all incidents where breathing apparatus (BA) sets are in use, it will be necessary for an entry control point (ECP) to be set up. At smaller incidents where only a few BA sets are in use (Stage I procedures are in use), the ECP will be under the control of a suitably qualified operative. The BAECO will communicate directly with the IC or with the sector commander if a sector has been instigated. The organisation of the ECP will follow standard national operational guidance, but in terms of the incident command processes, a BAECO should be monitoring communications with BA teams and with the sector/IC. A BAECO should regularly update sector/IC with information regarding:

- the number of BA teams deployed (and firefighting equipment in use) and their locations

- the progress made by BA teams

- the number of resources ready for deployment

- the resource requirements for the next 20/30 minutes
- any other urgent operational issues that may impact on the management of the incident.

Breathing apparatus entry control point supervisor (BAECPS)

When an incident gets more complicated and two entry control points and greater numbers of BA wearers are in use (Stage II procedures are in use), there is a need for a greater control co-supervision. Where this is the case, the IC (or sector commander) is required to organise emergency arrangements and establish communications with all other BAECOs at the incident. In addition, a breathing apparatus edge control point supervisor (BAECPS) with a suitable level of managerial authority and competence should be nominated by the IC or sector commander. This provides a greater degree of supervision and support for BA wearers and BAECOs.

The BAECPS will be given a briefing by the sector/operations/IC about the objectives and plan for operations using breathing apparatus.

The supervisor will then be responsible for:

- briefing BA teams and BAECOs prior to undertaking operations
- debriefing all BA wearers
- providing communications and updates from the control point to the sector/operations/commander
- liaising with the search co-ordinator to ensure a co-ordinated search is carried out and records of that search are maintained
- ensuring the provision of emergency teams and equipment
- supervising the communications for the ECP.

Support sector or functional roles

In order to manage the ICS, the IC relies on the support of others to ensure that events are managed effectively and safely for both the community and the firefighting teams they are responsible for. The command support function provides the IC with a supporting framework to ensure that the management of the incident is consistent, appropriate and joined up. The command support function may consist of a single person operating from the back of a fire engine (command point – indicated as such by having the only set of blue lights operating on the incident ground) or it may be specialist command support unit crewed by to six firefighters, managed by a command support officer, and can be itself supported by additional fire crews. The command support team (CST), command support officer (CSO) and the command support unit (CSU) are so essential to the safe operation of an incident command system that the next chapter will cover command support more extensively. It is through the CSO and the CST that support sectors/functional roles are co-ordinated and organised.

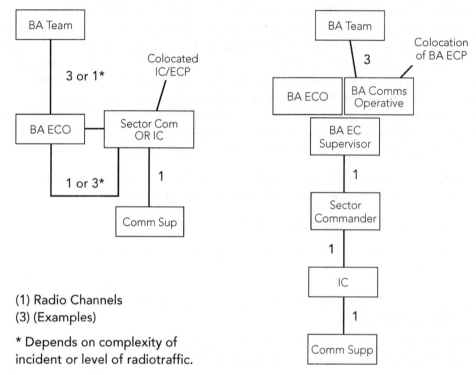

(1) Radio Channels
(3) (Examples)

* Depends on complexity of incident or level of radiotraffic.

Figure 8.26: Chain of command and communications network where BA operations stage I and II are in use

BA sector commander (BASC)

At large and complex incidents where BA is extensively used, a BA sector commander should be established to ensure that logistical support is available to ensure that resources do not become depleted, and to provide a greater degree of co-ordination between entry control points and BA supervisory teams across the incident. The BASC will be appropriately qualified and competent officer of sufficient managerial authority within the FRS. She or he may be supported by a number of assistants and be based either at a breathing apparatus support unit (BASU) or in an auxiliary unit.

The BASC has a number of responsibilities:

■ Establishing communications with all BAECPs, command support unit and logistical teams.

- Establishing an organisational structure chart for the BA function across the incident including the location and supervisor of each entry control point, the supervisor responsible for the rest and recuperation/welfare area and details of specific communications arrangements for the BA sector.

- Managing BA resources and logistics, including:
  - identifying and recording resource requirements for each BA entry control point
  - providing BA resources (equipment, firefighters, guidelines etc) to entry control points
  - ensuring reliefs arrive at the required ECPs at the required time.

- Establishing an area suitable for rehabilitation and rest of BA wearers.

- Maintaining records of equipment usage and wearers' details etc. as required by FRS policy.

Safety sector commander/safety officer (SSC/SO)

Ensuring a safe workplace is a key responsibility under the Health and Safety at Work Act (1974). In order to achieve this safe workplace, the IC may appoint an officer to undertake the role of safety officer to monitor and support those on the incident ground with command (or other) roles. The safety officer (SO) is responsible for monitoring the incident with the intention of identifying safety issues including the absence or incorrect wearing of personal protective equipment, dangerous structures, and smoke direction.

At larger incidents, where sectors and/or aspects of the incident demand greater level of control or observation, it may be necessary to point more than one safety officer and have the safety function managed by the safety sector commander (SSC).

Among other things, the SSC will be responsible for:

- managing and monitoring the safety sector function, which may include the appointment of sector safety officers (SO) and safety observers (SOb)

- reviewing and co-ordinating risk assessments from each sector SO and SOb and ensuring that control measures identified are implemented and co-ordinated across the incident

- organising communications within a safety sector in line with the incident command networks

- ensuring that the IC's understanding of safety issues within each sector is solid through personal observations and confirmation

- correcting health and safety issues when observed or reported
- co-ordinating the response to an accident in terms of investigation and reporting
- informing the command team of key issues identified.

Safety observer (SOb)

This is a firefighter or officer appointed to carry out a specific safety role at an incident. Reporting directly to the sector commander and indirectly to the SO/SSC. They may be asked to monitor fire conditions, take observations of building structure or pay particular attention to aspects of the incident which require close scrutiny (such as watching for the positioning of BA wearers within a building, ensuring they do not stray too far into the structure). This is a key role and has to be effectively briefed before deployment. If appointed, the SOb should not stand down unless ordered to by the SO or SSC, or properly relieved.

Box 8.5: Miscommunication – safety measures and safety observers

At a large town centre fire, a hot fire that required arduous work by BA wearers, it was decided that as well as an SO, a second safety officer should be appointed in an operational sector specifically to observe the activities of the BA teams due to the external temperature and humidity. The second SO was told to observe that the BA wearers did not proceed beyond a certain point due to the fire conditions. This additional control measure was applied for several hours. Upon change of shift, the relief of crews took place but the use of a second safety officer, which was a variation of normal policy and therefore not formally recognised, was omitted accidentally. A subsequent BA team entered the building and went beyond the previously demarcated point. They became disorientated and were overcome by heat exhaustion. One of the firefighters died from the effects of heat.

Lessons learned

When firefighters are being relieved it is essential that a full briefing is given to those relieving crews in their entirety. From the IC to the firefighter on the BA team, a full briefing including incident structure, objectives tasks, AND a safety briefing must always be given. If a safety sector commander has been appointed then they should brief the oncoming SSC or whoever is taking on the safety function as part of their role.

Additional safety measures undertaken that supplement existing procedures should be noted in the ARA and in the decision log, both of which should be read and understood by the incident commander, operations commander, command support officer AND safety sector commander.

Hazardous materials officer

At incidents where hazardous materials have been identified or are suspected, a HAZMAT specialist (sometimes categorised as 'Tactical Adviser' – see below) may be mobilised to assist the IC in dealing with these substances. Depending on the FRS, they may be designated as a 'hazmat officer', 'hazardous materials and environmental protection adviser' (HMEPA), 'hazardous materials environmental protection officer' (HMEPO). The designation 'Hazmat officer' will be used in this text. The hazmat officer has several roles in support of the IC in resolving the incident:

- Supporting the IC in determining the hazard areas and securing the area.

- Identifying hazardous materials using 'detection and identification of materials and equipment' (DIMM) teams and other specialist advisors.

- Undertaking a hazard and risk evaluation.

- Working with specialist advisers from other agencies including Public Health England (and other national equivalents), the Environment Agency (EA), National Chemical Emergency Centre (NCEC) and site specialists to identify options for resolving the incident.

- Advising the IC on the correct personal protective equipment (PPE) for firefighters to use.

- Utilising the best available information technology to identify the most appropriate response options.

- Ensuring that health, safety and welfare systems are in place for firefighters and other persons who may be affected by any hazardous materials. This includes the organisation of decontamination, medical support and evacuation in liaison with blue light and health services.

- Providing effective advice to the IC about legal and service matters regarding site decontamination, remediation and disposal of hazardous materials.

- Recording all relevant details to support enforcement action by other organisations including the Environment Agency, environmental health departments and Public Health England.

At more complex incidents, the hazmat officer may be supported by a decontamination officer and decontamination team. Resources may be provided from local, regional or national levels, including the capability for mass decontamination of members of the public. In some services, HAZMAT resources are provided remotely by other organisations and may take the form of an emergency telephone 'hot line'.

Decontamination officer

The decontamination officer is responsible for running the decontamination process under the supervision of the hazmat officer. It is likely that the decontamination officer is a watch or crew commander working at a station with specialist hazmat equipment. At larger or more complex incidents, this role may be carried out by a hazmat officer.

Logistics officer

At incidents where large quantities of equipment may be required, the use of a logistics officer will be valuable in managing the organisation, storage and transport of equipment across the incident ground. Reporting directly to the CSO, through the CSU, the logistics officer will:

- identify a suitable location for storing equipment so that it is readily accessible and can be dispatched to the relevant sector with these

- organise a team of firefighters to support the management and dispatch of equipment

- arrange suitable transportation to move equipment across the incident ground

- establish a communications network to ensure that requests for equipment are received quickly and that duplication is avoided

- ensure that command support are regularly updated with usage rates of equipment and provide an assessment of resource requirements over a prolonged period.

Marshalling officer (MO)

The marshalling officer is responsible for the establishment and management of a marshalling area. They will ensure that:

- a marshalling area is identified in consultation with the IC and the police

- the IC is informed when the marshalling area has been established, so that Fire Control can be advised of its location, and that all appliances not already in attendance and all additional appliances are redirected/directed to i

- in the case of a major incident the outer cordon RVP controller has been informed of the location of the marshalling area

- a communications link is established to the Command Support Unit

- appliance commanders are booked in attendance on arrival at the marshalling area (or RVP in the case of major Incidents)

- the CSU (booking-in point) is regularly appraised of the resources that are available

- appliances/crews are dispatched to the booking-in point at the request of the CSU

- crews remain with their appliances while at the marshalling area, and retain their nominal roll boards until required to book in

- the MO should be informed of any assistance message, and the additional resources that have been requested by the CSU

- the MO shall inform the logistics sector officer of any appliance that is available at the marshalling area without a crew.

When establishing a marshalling area, the MO should be cognisant of the following points:

- The establishment of the area will not affect the access and egress to the incident.

- The site should have adequate access for appliances.

- The site should have an adequate turning area.

- Ground conditions are likely to deteriorate with use and adverse weather conditions.

- The site should be capable of supporting the combined weight of appliances.

Fire investigation officer (FIO)

A specialist fire investigation officer may be mobilised to a fire as a direct request from the fireground or as part of a predetermined level of response. In the early stages of a fire, their role may be limited to observation and recording of details to enable them to carry out investigation once the operational phase of the event has been completed. Reporting directly to the CSO and command support, they may record the details of potential witnesses (FRS personnel, other emergency services or members of the public) and gather information from mobile data systems. Where the cause of the fire is suspected to be deliberate, they will liaise directly with the police to facilitate early investigation of a potential crime scene.

Fire safety officer (FSO)

Similarly, a fire safety officer may be mobilised to support an investigation into the cause and development of a fire with particular regard to potential failures of fire safety measures within the building. Where it is suspected that requirements of the Regulatory Reform (Fire Safety) Order (2005) have been breached, the FSO will liaise directly with the police and IC regarding the preservation and protection of evidence.

Incident monitoring officer (IMO)

Most services can deploy an officer to an incident with the role of monitoring the IC for the purposes of quality assurance, mentoring and coaching. This mobilisation is undertaken with the proviso that, should the incident require it, the IMO will take command. Circumstances under which this may occur include the potential size and complexity of the incident or circumstances that require an urgent change of command because the incident is a type of event that is beyond the current experience of the officer in development (or being monitored). This role is also known by other titles including 'active incident monitoring (AIM) officer', and some services use the term 'tactical adviser' (Tac AD), which can be confused with the role described below.

Tactical advisors (TacAds)

Tactical advisers are officers who have been given specialist training in various operational aspects of incident response. They will have specific references and may be available on a regional or national basis as well as within their own service. They are able to supplement the skills and knowledge that exists within services when specialist equipment or liaison with other agencies is required. They may be required to provide advice and support at any location, although normally they are used within their own (or 'host') service.

Tactical advisers should not take command of an incident, and legal responsibility for the management of the incident remains with the IC. They attend to provide the IC with advice and support to resolve the incident. They need to be aware of the IC's aims and objectives and the tactical plans developed to achieve them. Where necessary, advice given and decisions made based on the advice of TacAds should be recorded in the decision log along with the rationale and whether that advice is or is not followed.

At the most serious of operational incidents, there may be a requirement for multiple TacAds from all blue light services. It is of great importance that communications take place between them and that any advice developed collectively is communicated to the most senior person of each blue light service to ensure cohesion of strategy and tactics at the incident.

Interoperability is dealt with in Chapter 15.

Specialist equipment and techniques

The matters on which TacAds advise within host services includes the following:

- Water rescue.
- Mass decontamination.

- Bulk foam.
- Marauding terrorist attacks (MTA).

Other resources that are available, along with respective TacAds, on a national resource basis include:

- Chemical, biological radiological nuclear (explosive) (CBRNE), including detection, identification and monitoring (DIM) and mass decontamination.
- High-volume pumping units (HVPs).
- Urban search and rescue (USAR).
- Enhanced logistics support (ELS).

National inter-agency liaison officer (NILO)

A NILO is a specially trained and qualified officer who can advise and support ICs, police, medical and military personnel, and other government agencies on the FRS's operational capacity and capability to reduce risk and safely resolve incidents.

There are four main categories of incident at which the NILO may be involved:

- Conventional and CBRNE terrorism.
- Major incidents.
- Complex or protracted police-led incidents.
- Spontaneous and planned serious public disorder.

Non-fire and rescue specialist personnel

Besides blue-light emergency responders, there are likely to be a range of personnel from other organisations and agencies that can support an IC in resolving an incident. Apart from the occupier of any premises, an IC can expect to see attendees from all category 1 responders (as defined under the Civil Contingencies Act (2004)), especially those representing the health services, the Environment Agency, Public Health England and the local authority. Category 2 responders may also attend, dependent upon the nature of the incident, including representatives of utilities, transport providers (road, rail, air transport, maritime) and organisations such as the Highways Agency and Network Rail.

All non-emergency service personnel are required to 'book in' at the command point or command support unit, register their name and vehicle, and give contact details before being issued with an identification tabard. Normally, liaising with these personnel will be the responsibility of an appropriately nominated officer. For example, the Environment Agency may have a hazmat officer allocated to them for liaison purposes. The marshalling officer may be required to liaise with the police to organise access and egress routes to the incident. The CSO will co-ordinate this work and ensure that all information is collated and provided to the IC to ensure a full understanding of issues supplementary to the operational activities within sectors.

Summary

You should now understand the following concepts, techniques and practicalities discussed in this chapter:

How an incident ground is organised and its key components.

The concept of spans of control and the minimum and maximum number under dynamic and stable conditions

The management of growing incidents including the use of sectors, alternative command structures and the growth of command structures to meet the increasing demands of an incident. Using novel structures to meet the operational needs of an incident.

The use of sectorisation for various specific incidents including high rise, basement, transportation and wildfire incidents.

The use of cordons as a control measure for the safety and operational efficiency of firefighters.

The importance of hazard areas and the determination of incident tactical mode and associated messages and information sent to fire control. The changing of tactical mode and the use of tactical withdrawal and emergency evacuation to maintain a safe working environment.

The use of an incident 'reset' when the risk to firefighters is too high or conditions have deteriorated.

An understanding of the key command and support roles used at an operational incident including the incident commander, operations commander, sector commander, safety officer, command support, BA entry control supervisor and operative, mass decontamination and tactical advisors.

Chapter 9: Organising the incident ground – the command support function

Introduction

In order for an IC to function effectively, it is necessary for them to be supported by a range of personnel and equipment that comprises the command support function. For smaller incidents, one firefighter using a command support pack may suffice for managing up to 20 or 25 firefighters plus supporting appliances. Beyond this, when incidents get to six appliances or more, it is generally the case that a specialist vehicle, known as a command support unit, plus crew and specialist officers, is mobilised to support the incident command structure.

The incident framework depends on effective command support function for its safe and effective operation. This chapter looks at this key aspect the command support function from a simple two or three pump incident to large and complex major fires and other incidents. It considers its implementation, what equipment it should require, how the functioning of the unit can be managed effectively, and how it supports safe systems of work and command effectiveness.

The first attendance: incident command point

On arrival at an incident, with only one or two fire engines and up to nine firefighters, setting up command support function is not usually the immediate priority of an IC. It is important, however, that command support in the form of a command point is set up as soon as circumstances and resources allow (initially in the cab space of an appliance or the back of a pump; ideally a vehicle not being used as the primary pumping appliance). The vehicle being used as the command point should be the only one on the incident ground with its blue hazard lights flashing and may be equipped with a red flashing light (or occasionally a red and white chequered flag) designating its role as command point.

Command support should always be undertaken by a suitably qualified and competent person. At smaller incidents the command support function can be carried out by a single firefighter who will be responsible for:

■ collecting nominal roll boards

■ sending messages from the fire ground on behalf of the IC

■ briefing oncoming crews as to the nature of the incident, safety issues etc Before sending them to the IC for tasking

■ preparing incident sketches to assist with the deployment of oncoming crews

■ when time permits, conducting a preliminary assessment of risks on the incident ground.

The equipment required to support this function at a small incident is minimal and usually consists of radio access to fire control, a fire ground radio, a pad of paper and pens, plus associated aide memoires and guides for the function of command support. Where command support packs are available, they should be used wherever possible because it will develop knowledge and familiarity with the pack and its role within command support. The firefighter responsible for the command point should also wear the command support tabard, kept on an appliance for this purpose. At a growing incident she/he should continue to carry out the command point operative role until relieved when the command support unit starts operating.

Where the incident is likely to grow or is prolonged, it is likely that a more detailed recording process and systematic command support functions will be required. While the equipment on appliances varies with each FRS, command support packs (colloquially known as 'pizza packs') should be provided on appliances for use by those first attending an incident. All personnel should be familiar with the equipment carried in the packs through use at incident and training exercises. While not definitive, the items listed below should be included within the command support pack:

■ Command support tabards.

■ Resources for checking nominated roles (for both appliances and officers).

■ Dry wipe or chinagraph board for recording significant incident details.

■ Pre-engraved command structures templates and site plans.

■ Message pads.

- Incident command log.
- Risk assessment forms.

The command support pack should be used until all the information has been transferred from the command points to the command support unit and a formal handover of command has been undertaken and notified both to fire control and incident ground.

The larger incident: command support units

When an incident grows beyond a certain threshold, additional command support resources are mobilised. Dependent upon an individual service's policy, specialist dedicated vehicles, command support units (CSU), are mobilised to provide additional control facilities for the IC. In some services this may take the form of a two-stage support process whereby, for incidents requiring four or more pumps, a forward command unit (FCU) will be established, and a larger CSU will be dispatched when the number of pumps required exceeds six, eight or ten (depending upon the service's policy). Services with just one type of command unit will usually mobilise it at a designated threshold of between six and ten pumps.

Forward command units (FCU)

These units are normally mounted on a large sports utility vehicle (SUV) chassis and are equipped with rudimentary command facilities, which may include communications equipment including GSM phones (Global System for Mobile), command boards, whiteboards, laptop computers with a Wi-Fi link and MDTs with printers. These units are crewed by two or three staff who are competent in the use of equipment and have a good understanding of incident command systems and processes. The relatively rapid attendance of these units at an incident will help support safe systems of work at an escalating event.

One downside of forward command units is that, as an incident continues to grow and the command support units are mobilised, there may be an additional transfer of command – the first transfer taking place from the command point to the forward command unit, and the second transfer between the forward command unit and the command support unit. These transfers need to be carefully managed, particularly when a command support unit is mobilised within a very short time of the forward command unit, with both arriving virtually simultaneously.

The functions of a forward command unit normally falls between those of a command point and the full capabilities of the command support unit, although under certain circumstances they would be able to sustain support for large incidents over an extended period.

Figure 9.2: Side and rear views of a typical forward command unit (FCU)

Command support unit (CSU)

Command support units are normally mobilised to incidents above a certain number of deployed resources, which will depend on the nature of the incident and on the service, and may consist of combinations of whole time and on-call firefighters as well as fire control staff. Equipment on both the FCU and the CSU is technically complex and requires regular use and exercising to keep up with the skills required to manage the unit effectively. Staff must therefore maintain their competency on the equipment and their understanding of the processes involved in incident command.

Equipment on CSUs should include, as a minimum:

- An array of communications equipment to enable contact to be established and maintained with fire control, other agencies, FRS incident support rooms, and direct links to government departments in the case of major incidents. This equipment includes a range of mobile telephone communications, GSM phones, satellite communications, two-way radios and fire ground radios.

- ICT systems capable of replicating all digital information available on mobile data terminals, SSRI information, digital maps including special applications for water resources, road networks, meteorological information and satellite TV links.

- Printers and photocopiers.

- Batteries for radios, spare radios and GSM phones.

- Charging facilities for phones and computers etc.

- Video surveillance cameras and digital recording devices.

- A variety of whiteboards pre-engraved to record the attendance and deployment of crews, hazards and risks on the incident ground, incident command structure, messages and the time line of activities required and completed. Most of these functions can be replicated using IT software systems but the provision of whiteboards gives the CSU a resilience that enables it to function even where the power supply is lost and IT systems have crashed. Where possible, tripods and stands should be provided to allow relocation of the command support unit in the event of the vehicle becoming untenable.

Figure 9.3: Examples of a CSU

Box 9.1: Changing command support

At a large fire in a high-rise building, the initial command point was set up on the pump from the first attendance at around 2014hrs. A special equipment unit (SEU) with command support facilities arrived four minutes later and formally notified fire control that it had taken over as incident command point at 2033 (19 minutes after the first attendance command point was set up). A further change of command support occurred when the initial command support unit (ICSU) took over at 2119hrs but only a limited transfer of information was carried out from the SEU, and so a comprehensive understanding of the situation by the ICSU team was not likely. A larger command support unit (CSU) arrived in attendance at 2115hrs and at 2131hrs (13 minutes later) took the formal command support function for the incident. Again, only a limited transfer of information occurred from the initial command support unit. Contact had been lost with two firefighters and a BA emergency had been declared at 2108hrs.

So, during this period of intense activity and high emotional stress, incident command had been transferred three times, and each time information transfers had been incomplete.

Lessons learned

- When transferring information from command point to command support, ensure the information is complete and accurate.

- Use personnel who have been involved carrying out command support to facilitate the transfer and remain with the oncoming command support unit to help 'bed in' the new staff and assist them in making the transfer seamless.

- It can be reasonably anticipated that, when it is likely that a major command support vehicle will be used and will be available within a short time, it may be worthwhile delaying transfer of command support to an intermediate unit. This reduces the potential for losing information during the transfer from unit to unit.

- It may be worth delaying the transfer of command unit functions when activities are of a high intensity and when the transfer process may interfere with that operational activity. When the incident has settled down to a relatively steady-state, then the transfer of command support may take place.

Command support units normally provide a conferencing facility that allows discussion between the command team and other agencies in relative quiet. It may also be used as a quiet area to allow the IC to consider options and strategies without disruption. This area may take the form of separate room within the command support unit or in a separate inflatable tent or integral awning. When the IC or others are in the conference area, it is important that they are not excluded from events on the incident ground and it is the role of the CSO or CSU team leader to act as a conduit for key information.

Figure 9.4: Conference area of a command unit

Siting of the CSU

The optimal location for a command support unit or for control unit will depend on the nature and extent of the incident. Factors to be borne in mind when siting the unit include:

- Proximity to incident – it should be near enough to be able to command and control the incident while remaining remote enough not to be affected by incident ground activities. Ideally the CSU should be sited where it can help control access and egress of resources onto the incident ground.

- The CSU may remain on site for a considerable time so it's important not to park the unit on an area that may become inundated with water, run-off or contaminant. Parking on soft surfaces may initially seem appropriate, but when saturated with water it can cause vehicles to become unstable and unbalanced, which could in turn affect communications.

- Avoid parking by tall buildings or structures with large amounts of steel framework, which affect communication signals.

- Consider creating an all-services 'command village' where resources can be pooled (welfare facilities including feeding and refreshment), access to senior commanders is readily available, and security issues for equipment (and staff) are simplified.

- It is important that when attending a multiagency incident where all of the agencies have deployed their command units, that consideration is given to the specific siting of these units. It is possible with both satellite and radio-based communications that interference may occur which distorts and sometimes neutralises signals. Where command villages are setup, this effect may be multiplied. Is therefore important that exercises and trials utilising these units together will enable guidance to be produced regarding the siting of units and help operators identify best practice.

The roles and responsibilities of the CSU

Each service will have its own expectations of specific functions which the FCU and CSU undertake. As a minimum they should include requirements set out below:

- To provide first contact point for all attending appliances and officers.

- To maintain a 'live' record of resources in attendance at the incident.

- To maintain a record of all attendances at the incident.

- To provide safety briefings and operational briefings to all arriving appliance crews and officers.

- To create and update an incident ground organisational chart and record sector identifications.

- To deploy resources to a sector, function or marshalling area as instructed by the IC or CSO. They must record and maintain records of all deployments.

- To liaise with the marshalling officer and police to organise access and egress routes, holding areas and rendezvous locations.

- To liaise with other agencies as necessary, including booking-in and supervision of their staff, managing the media etc.

- To support the safety officer/safety sector commander to record the outcomes of the analytical risk assessment (ARA), any reviews and any operational decisions or actions taken as a result of the review.

- To maintain the incident decision log and record decisions and rationale for all decisions as instructed by the IC or CSO. Prompt the IC and CSO when decisions need to be recorded.

- To maintain a time line for ensuring key activities (ARA reviews, declarations of tactical mode, pre-arranged meetings with others, informative messages) take place when required.

- To operate the main-scheme radio link to the brigade control and to log all main-scheme radio communications.

- To assist the IC in liaising with other agencies by providing logistical and support functions for meetings.

- To work with fire control and the incident support centre (when running) to facilitate the smooth relief of appliances, officers and support staff.

- To arrange supervision of and liaison with incident ground welfare facilities providers.

- To aid visits by dignitaries, councillors, etc.

- To raise issues with the IC, operations commander and command support officer in order to improve the management of the incident.

Figure 9.5 : A team meeting in the conference area

The command support team

Command teams

In an ideal environment, the IC will have a team with them composed of individuals with the correct balance of skills, with whom they have worked previously and have a good working relationship. Having practised as a team in exercises and incidents, such a team will be able to work together effectively to set up quickly and get working. Ideally, the people picked for the roles within the team will be well-suited in terms of knowledge, skills and attitude, but with time being a critical factor, this may not always be achievable. It is more likely, in fact, that the IC's supporting team will not have worked together previously and will arrive at an incident in a discontinuous stream.

The team may therefore have to develop on an ad hoc basis with positions being filled by those available rather than those who may be best-suited to the actual role. If possible, the IC should plan the structure of the team based on the skill sets of those who may be known to be attending the incident (this information can be gained from fire control or the IC may be part of a standard mobilising protocol under which pre-determined officers will be mobilised together). Certain officers may be mobilised to carry out particular roles, such as fire investigation or command support. This does not preclude the IC reassigning them to other duties if circumstances dictate (e.g. if an officer designated as fire safety officer arrives before a command support officer, it could be prudent to assign that individual to the command support role).

A well-run CSU will be the focus of the incident with the command team consisting of the IC, an operations commander (where appointed) and the command support officer, together with CSU supervisor and team. The CSU supervisor is normally a watch manager or crew commander with specialist knowledge of the systems and operating procedures to be used within the unit. The operators that make-up the crew should likewise be competent in the operation of equipment on the CSU and have a good understanding of incident command systems and their role in managing it on the incident ground. The team should be briefed about the status of the incident as a whole wherever possible, as this will aid shared situational awareness. Each member should be sufficiently motivated to work within their own allocated area of responsibility with a minimum of supervision and should proactively be seeking to support the IC and the command team. The closer the command team and the IC work, the more effective the ICS.

Responsibilities of the command support officer (CSO)

Overseeing the command support function is normally the responsibility of a dedicated command support officer (CSO). The individual allocated this role should be familiar with the workings of the CSU and the underpinning incident command system and its processes and procedures. In some services, officers receive specific training on the role of CSO and are allocated to incidents specifically to carry it out. Other services take the view that undertaking the role of CSO is an extension of the tactical incident commander role.

A command support officer is likely to arrive at an incident before the CSU and is in an ideal position to supervise the siting of the vehicle. While the CSU is setting up, the CSO should endeavour to obtain a good understanding of the incident including details of operational strategy and plans, resources in attendance, the incident command structure and other information that can be used to support the introduction of the CSU. As soon as the CSU is at a stage where it is ready to assume the command support function, the CSO will make the decision to make the transfer, bearing in mind the considerations mentioned in Box 9.1 on p171. As the officer responsible for the effective management of the command support unit, the CSO must ensure that the following activities are properly implemented, supervised and organised:

- Establishing command of and supervising the CSU team. Provide a briefing to the team and allocate roles and areas of responsibility to each member of the team. As a minimum, the roles required in the CSU are as follows:

    - a radio operator for communications with fire control

    - a radio operator to manage incident ground communications

    - an individual tasked with maintaining incident ground resources and operational structure charts, deployment of staff and roles of sectors and crews

    - an individual should be nominated to carry out the role of completing the decision log

    - the supervisor for ensuring activities are carried out efficiently.

- Assessing the capabilities and capacity of the team and ensuring it is effectively resourced. Where this is not the case then additional resources should be requested in the form of a supporting appliance to enable staff to carry out ancillary activities such as drawing plans, acting as runners and providing cordon control for access to the CSU.

- Overseeing and helping co-ordinate the handover of responsibilities from the initial command support location. As mentioned above, the transfer from the control point to the CSU often causes problems and delays due to the ineffective management of this transfer. An effective CSO will not rush the transfer as this often leads to incorrect or insufficient information being sent to the CSU with all the complications that may arise.

- While the command support unit team will be responsible for physically setting up the main scheme radio link and allocating fire ground radio channels, the CSO must ensure that this is completed in line with existing arrangements.

- On behalf of the IC, the CSO acts as the line manager for functional sectors including water and safety, as well as being initial liaison with other emergency services and other category 1 and category 2 responders.

Problems associated with CSUs and the command support function

The IC should always be aware of potential problems with the ICS in respect of equipment and systems. There have been a number of high-profile incidents where accidents on the incident ground could have been due to problems with command and communications. Potential problems include:

- A failure of incident ground radios due to a lack of batteries or recharging facilities.

- An overload of mainstream radios due to the large-scale nature and demands of an incident.

- An overly slow handover of command from command points to the command support unit.

- Multiple changes of command due to a sequence of increasingly senior officers arriving at the incident taking charge, leading to confusion within the CSU with several commanders believing they are in charge simultaneously.

- Overcrowding within the command support unit. Many people may want to enter the CSU to gain an overview of the incident. Entry to interior of the CSU should be prohibited and controlled, usually by the supervisor of the CSU crew.

- Large numbers within the cabin of the CSU increases background noise and the potential for misunderstanding the messages being received from the fire ground, from control, or from other personnel within the unit.

- Failure to challenge decisions being made. Staff should be encouraged to speak up where they feel an incorrect decision has been made as a result of information the IC may not be aware of.

Summary

You should now understand the following concepts, techniques and practicalities discussed in this chapter:

The command support facilities available on the first and subsequent attendances including the command point and command support unit.

An understanding of the practicalities of the siting and use of the CSU.

The types and functions of command support and those who have the responsibility of managing and running the command support unit.

Chapter 10: Safety on the incident ground (part 1)

Introduction

All firefighters, upon signing the contract for employment, accept implicitly that there is a risk that they may be injured or worse in the execution of their duties. Those who seek and obtain promotion have an even greater responsibility in that they are in a position of authority and command decisions they make can result in the injury or worse of other firefighters. It is incumbent on all firefighters, ICs and commanders to monitor and protect the safety and health of their colleagues.

Where this risk is most acute is naturally at operational incidents where, despite the relatively predictable behaviour of fires, buildings and vehicles, there is a possibility of unforeseen and unpredictable events. It is important for firefighters to understand the principles of health and safety on the fire ground to ensure their safety, the safety of their colleagues and the people who pay their salary – the community. The failure to effectively manage incident ground health and safety can not only result in injured firefighters, but also represents a waste of resources, potential loss of community assets and of members of the community themselves.

A systematic approach to health and safety on the incident ground and an understanding of its principles and its practical application means that firefighters are safer and ICs can have confidence in the competence of their teams. The principles involved include an understanding of the 'safe person concept', which means that individuals are selected, trained and promoted on the basis they are competent to work within a hazardous environment. It presumes effective training regimes and robust assessment processes, and recording practices that deliver a safe and effective firefighter.

Some definitions and concepts

Remember, it is essential that firefighters understand the distinction between the terms 'risk' and 'hazard': a hazard is something that has the potential to cause harm. Practically, this may be a building wall that has been damaged by fire and in danger of collapse. A 'risk' is a measure of the likelihood of a hazard being realised and actually causing harm. In this example, it would be the wall falling and causing either physical or psychological harm, damage or loss to property, environmental or economic losses. Taking the example further, the IC could put in place a control measure that would reduce the risk that the collapsing wall would cause harm. The control measures could include cordoning off a collapse zone area (one and a half times the height of the building), setting up monitors for putting water onto the fire and placing safety observers to ensure that no one breaches the cordon.

The identification of the hazards on incident grounds and the relative risks is assessed by a number of processes that introduce such control measures that minimise or reduce the risks to as low a level as can be reasonably achieved. Three types of risk assessment – dynamic, analytical and personal – are discussed in this chapter.

The 'safe person' on the incident ground

The Health and Safety at Work Act (1974) places a duty on employers to ensure the health, safety and welfare of their employees. It is also incumbent on the individual employee to take reasonable care for their own health and safety and that of others who may be affected by their actions. They also have a duty to co-operate with their employer to meet their requirements under the Act. While the organisational responsibilities are dealt with in a host of other guidance, and elsewhere in this book, it is important that individuals on the incident ground including the IC are aware of their own individual responsibilities towards others.

Training and experience will help firefighters to manage and mitigate potential hazards and risks on the incident ground, but there are a suite of elements that everyone should be aware of. The first is an assumption of the competency of the 'safe person' – i.e. they should be capable of carrying out the tasks allocated to them by the IC or subordinate commander in a safe and effective manner. It is assumed that they will work sensibly and responsibly using the command and control systems that have been developed for that specific incident and of which they have knowledge and understanding. Equally, there is an assumption on their part that those directing the activities are also competent in their role of supervising tasks as well as monitoring and supporting them.

It is incumbent upon the individual that they are cognisant of their own capabilities and limitations in terms of knowledge, experience and physical abilities. The individual should also be confident and assertive enough to raise awareness of their own limitations when given a task they feel is beyond them, and request additional support or seek alternative methods of achieving the objective. For example, a member of a BA team who feels unfit to undertake an arduous operational activity should not hesitate to advise the supervisor of the fact as soon as possible so that alternative steps can be taken.

A key skill for a trained firefighter is to have an acute awareness of the situation and environment in which they are operating. They have a responsibility to themselves and to the team. They are required to be vigilant for emerging hazards, and be prepared to take appropriate action to mitigate any emerging risks. It is also their duty to share knowledge of emerging risks and changes in operational conditions with supervisors and teams, thus helping to provide the IC with a sound understanding of an incident, both outside the hazard area and within.

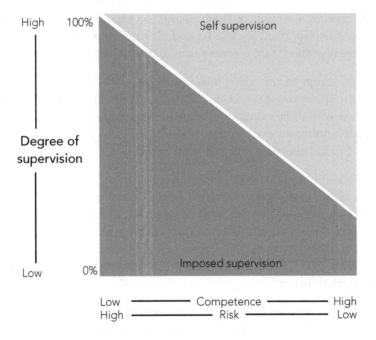

Figure 10.1: Supervisory requirements in risky environments and varying staff competence adapted from HSE

Ensuring firefighter safety is also a key role of every member of the supervisory team and incident management structure all the way to IC, who is legally responsible for activities on the incident ground. All supervisors have a duty to ensure the safety of those within their charge and continually monitor crews and teams, making sure that safety practices are adhered to (visors in the correct position, flash hoods appropriately positioned around BA masks, ladders footed, dust masks donned etc.), and that bad practice is immediately corrected. Where bad practices have been identified, the supervisor should raise this as a serious matter through the debrief process or by using a formal health and safety reporting mechanism to identify whether this is a 'one-off' event or a systemic issue across the organisation.

Human factors

The UK Health and Safety Executive define human factors as, *'environmental, organisational job factors and human and individual characteristics which influence behaviour at work in a way that can affect health and safety'*. Interrelationships between job, individual and the organisation, can be managed effectively and it is important for the IC and all firefighters to recognise the potential for harm if these factors are not managed effectively.

Each individual brings to the workplace their own skills, knowledge and attitude. Personalities and interactions with work colleagues, their strengths and weaknesses, can all be complex, varied and either positive and negative. The influence of personality on our working practices may be positive or negative and can be difficult to change, whereas skills and attitudes can be improved.

The tasks involved in a job should, ideally, be designed with the human limitations in mind, and the work should be designed ergonomically. This includes matching physical and mental processes and capabilities (including information assessment and decision making capabilities, and understanding of the task and risks) to the needs of the task. It is the mismatch between the requirements of a job and the capabilities of an individual that can cause the potential for human error.

The final category of 'human factor' is the influence of the organisation on individual and group behaviour. The impact of organisational attitudes on its health and safety culture can be significant, and an unreasonably high tolerance of risk within an organisation will influence its employees, leading to a widespread culture of risk acceptance. A strong safety culture, on the other hand, can produce the opposite effect, where risks are kept in proportion and individuals work together to reduce risks and mitigate the impact of activities on individuals and the service.

Risk assessments

Risk assessment is the determination of a qualitative or quantitative estimate of risk in relation to a recognised hazard. Assessment comprises two elements – the severity of impact and the probability that it will occur. During emergency phases of unplanned incidents, the situation and hazards are often less predictable than those in planned activities. If hazards and risks are predictable, standard operating procedures, carried out by trained and competent firefighters, should be adequate to deal with the situation. Where the situation is dynamic and continually evolving, ongoing risk assessments must take place.

This process has been termed a dynamic risk assessment and has been defined as, *'the continuous assessment of risk in the rapidly changing circumstances of an operational incident, in order to implement control measures necessary to ensure an acceptable level of safety'*. As soon as time permits and when resources are sufficient, an analytical risk assessment should be undertaken. This is a more structured process, carried out by a competent individual, and it supplements the dynamic risk assessment with a more careful analysis of risks and hazards. Both are addressed in more detail below.

Dynamic risk assessment (DRA)

This is an assessment of hazards and risks made in a rapidly changing environment. It helps to inform the IC's plan and will assist in determining the tactical mode employed. The DRA should be carried out continuously by all those on the incident ground. Notifying fire control of the dynamic risk assessment outcome is achieved by sending a message stating the tactical mode in use.

Relating this to the decision control process (DCP), the DRA initially takes place within both the SITUATION and PLAN stages (Chapter 7) when assessing the information available, and deciding on tactics and control measures. The process of carrying out a dynamic risk assessment follows a number of sequential steps and is an iterative processes.

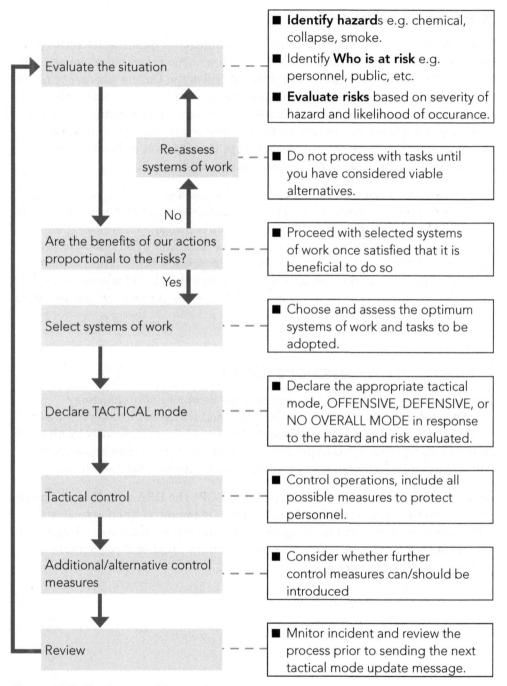

Figure 10.2: The dynamic risk assessment process
(Adapted from NOG)

Evaluation of the situation

The first task of the IC attending an event is to gather information about the incident, the availability of resources, and the potential risks and hazards faced by responding firefighters. Information gathered en route and on arrival from people, combined with visual cues, will assist this process. The risks identified may include the fire itself, whether any chemicals are involved in the fire, the likelihood of structures collapsing, etc.

The people involved in the incident who may be affected by hazards also need to be considered: firefighters, members of the public, occupiers and other emergency service workers. A judgement then needs to be made on the likelihood of hazards occurring and the risks they pose. In reality, this can be difficult as there will be unknowns at an event or fire which the IC is unlikely to be aware of at the time. Nonetheless, it is necessary to make a 'first stab', or educated guess, based upon current knowledge and professional judgement.

Benefits versus risks

Having decided a course of action following the initial use of the decision control process to determine the strategic objectives, the IC must consider what actions can be undertaken and what the likely benefits of those actions are going to be. It is necessary to carry out a consideration of the associated risks with that course of action, or to determine whether the risks are proportional to the likely benefits of success. If the judgement is that the likely risks are greatly outbalanced by the benefits of success, then the action should proceed as intended. For example, at a simple kitchen fire in a two storey house, with a person trapped in a bedroom, a course of action involving a simultaneous attack on the fire with a jet downstairs and a rescue via the internal staircase using BA crews and a hose reel jet may have benefits that outweigh the risks. The fire is likely to be contained by the jet on the ground floor, the casualty is likely to be rescued by the firefighters from the first floor, and both crews are being protected by jets and not likely to succumb to injury.

Conversely, continuing with this example, if the ground floor had a fully involved fire with heavy smoke escaping from the windows at the first floor level, then the situation would be very different. The risk of injury to firefighters has increased and it is likely that the chances of survivability of any casualties in the first-floor bedroom would be slight. Committing firefighters to a rescue without additional support and resources such as an emergency crew, additional covering jets or an aerial ladder platform, may place them at an unacceptably high level of risk. In this case the judgement is likely to be made that the risks facing firefighters as a result of the action is greater than the benefits of success. As a result, the IC must reconsider the actions that have been proposed, re-evaluate the risks and develop a different plan of action.

Select the safe system of work (SSoW)

Following evaluation of the situation and having carried out the risk-benefit analysis, it is then necessary to select appropriate SSoWs which will be required to put into effect the action plan. This may include fundamental safe systems, such as BA Stage I for crews entering the building, gas tight suits (for chemical incidents), operational equipment such as compressed air foam systems (CAFS) or hose reel jets, and safety systems such as observers or safety sectors being introduced. Additional SSoWs could include the use of thermal imaging cameras, BA communication systems and the nomination of a search co-ordinator.

Declare the tactical mode

Having carried out the preceding steps, which form the initial risk assessment, the IC is then in a position to formally declare the tactical mode of operations – 'offensive', 'defensive' or 'no overall mode' (see Chapter 8). This time stamps the risk assessment by transmitting it to fire control, where it will be recorded and stored.

Implementing tactical controls

The use of radios, effective briefing and debriefing, and the introduction of sectors and safety officers will be used to improve supervision, accountability and control over staff during operational activities. It is important during this operational phase that all staff feedback to the IC, reporting the conditions on the incident ground as this will help identify what other control measures may need to be added to ensure safety of crews and members of the public.

Consider additional or alternative control measures

When the incident, resources or hazards change, it may be possible to introduce additional or alternative control measures. For example, when a CAFS or ultra high pressure jet equipped appliance arrives at an incident, it may be possible to use foam or water mist as the extinguishing medium rather than jets.

Review

As with all risk management processes, dynamic risk assessment is an iterative activity requiring reviews on a periodic basis. In this case, all elements of the review process need to be confirmed or amended as necessary and actions taken as required. In this way, changes to the incident, the level of resources and the impact of service activity can be factored into the tactical decision-making process. The incident will be reviewed through continual feedback, and new information from crews, sectors, safety observers, neighbours etc, as well as the monitoring of the achievement of the objectives against the plan.

The DRA process can be quickly implemented but could at times be distorted by the dynamics, moral pressure and immediacy of the situation the IC is confronted with, and it is possible that some vital information may be missed, not available or not considered. This can result in unrecognised hazards leading to unassessed risks, and hence to limited controls put in place. This could be caused by tunnel vision, stress and fatigue, distraction or omission and could result in accident, injury or additional property loss.

The advantages of dynamic risk assessments are:

- They are quick to complete and provide an initial understanding of risks on an incident ground, and they help avoid the potential for crews to undertake instinctive activity without consideration of the incident or specific circumstances they are dealing with.

- It recognises the dynamic nature of the incident ground and is a quick and ready evaluation of risks, actions to be taken and control measures to put in place.

- It is a continuous process and evolves with the development of the incident itself ensuring currency of activities and actions being managed on the incident ground.

- It provides a template that allows quick decisions to be made on the deployment and allocation of resources across an incident and also helps identify what supporting measures, including resources, are required to develop a robust and safe system of work, balancing risks and benefits.

There are a number of disadvantages to the process that are primarily caused by the speed at which the assessment is undertaken. These include:

- The DRA process does not require a detailed list of hazards or calculated risks on the incident ground. It is very much an overview of the situation at that time. For the IC in first attendance, time is of the essence and action is required. Anything that delays taking that first action may have a significant impact on casualties, the spread of the fire or environmental damage, for example, and so it is imperative that actions are taken as urgently as possible. This means that the dynamic risk assessment must itself be carried out rapidly and it is possible that some hazards will be missed, some risks not quantified and some key aspects of the incident overlooked.

- There is no formal recording mechanisms for the outcomes of a dynamic risk assessment, other than the passing of a message to firefighters on the incident ground and as a message to fire control. Unfortunately, the lack of recording of the reasons behind the tactical declaration can mean that, for post-incident debriefs, information is missing and decisions taken may not easily be remembered some weeks after the event.

While the DRA is a first attempt at understanding the initial risks on an incident ground, it is this assessment that also provides the basis for a more systematic and controlled analysis of hazards and risks that should follow – the analytical risk assessment.

Analytical risk assessment (ARA)

As soon as it is practical, the IC should arrange for an analytical risk assessment to be undertaken by a suitably qualified individual. The ARA is a systematic recording of risks and hazards on the fire ground using a risk analysis tool, most often a matrix-based chart. Essentially, an ARA records:

- a list of hazards
- probable risks associated with those hazards, and to whom
- a list of control measures already in place
- a list of additional control measures to reduce risks that are identified as being moderate, high or very high, with the aim of reducing those risks to a tolerable level
- confirmation that the current tactical mode is appropriate.

Hazards

The incident ground is a hazardous environment, but not all hazards need to be recorded: recording should be limited to the 'significant risks'. Slips, trips and falls all regularly appear as hazards on ARA records but they are usually superfluous to any serious assessment as they are naturally occurring hazards that should be taken, as a fact of life. If, however, there are aspects of the incident ground which increase the likelihood of a slip or trip or fall occurring (such as adverse weather conditions, snow and ice, a sloping embankment near a railway track, soft ground etc.), then these everyday hazards should be recorded and assessed for increased risk. Any control measures already in place should be taken as part of the risk reduction measures and recorded on the ARA form.

Risk rating

The severity of the hazard in terms of injury or property damage is then assessed using a standard numerical rating of 1 to 5, which measures severity between negligible (1), and catastrophic (5). The severity of the hazard is measured with existing control measures put in place. The likelihood of the hazard occurring are also given a numerical rating of 1-5, with 1 being 'very unlikely' and 5 is 'certain'.

ARA Should be carried out as soon as time or resources permit

Name: Smith	Incident Address:	Date: 15/03/16
Rank/Role: AGC	The Croft, High Street, Horley	Time of ARA: 0300
Service No: 2121		Whole Incident: Yes/No
Incident No: 9999	Sector Name/Number: 3	Sheet....... Of....... ARA No (1, 2, 3 etc) 2

Hazards (Write details)	Risk Group/s	Existing Control Measures	Level of Risk	Additional Control Measures	In Place
Collapsed Building	A, D	Cordon - taped off. PPE	Tolerable	Minimal personnel	Y
Struck by object	A, D	Cordon - taped off. PPE	Tolerable	Safety Observer	Y
Uneven and slippery surfaces	A, D	Lighting and designated safe routes	Tolerable		
Poor Lighting	A, D	Portable lighting	Minimal		

Do any sector activities/hazards directly affect other sectors or agencies? Yes/No   If Yes have they been informed? Yes/No

Comments: Non-Saveable property. TL in operational in sector 1. Operational activities will result in water and debris falling into Sector 3.

Tactical Mode (please tick)	Offensive	Defensive ✔	Name & signature of IC/OC/SC: GC Jones

Time handed to command Support: 0320

ARA Review	Time: 0350	Reviewer: AGC Smith	IC/OC/SC Signature: AJ
	Time:	Reviewer:	IC/OC/SC Signature:
	Time:	Reviewer:	IC/OC/SC Signature:

Figure 10.3: An example of an Analytical Risk Assessment Form complete with 5x5 risk matrix
(Reproduced with permission of Surrey Fire and Rescue Service)

1:Hazard ID		2: Control Measure	
This list is not exhaustive		This list is not exhaustive	
Access/Egress	✔	Additional Resources	
Acetylene		Ambulance Assistance	
Animals		BA procedures	
Asbestos (High Hazard)		Breathing Apperatus	
Asbestos (Lower Hazard)		Covering Jets	
Biological		Crew Briefings	✔
Building Construction		Damping Down	
Chemicals		Decontamination	
Collapsed Buildings	✔	Distance	
Confined Space		Dust masks	
Disease		Ear Protection	
Electricity		Evacuation	
Environmenal Hazard		Eye Protection	
Fall from Height		Gas tight Suits	
Fatigue		Hazardous Materials Advisor	
Fire/Explosion		Hi Viz Clothing	
Gas, Mains/LPG		Isolation	
Insufficient PPE		Life Jacket	
Machinery		Lighting	✔
Manual Handling		Limit Exposure	
Moving Vehicles		Minimum Crew Numbers	✔
Noise		Police Assistance	
Poor Housekeeping		PPE	✔
Poor Lighting	✔	PRPS	
Pressure Systems		Repositionings of Appliances	
Radiation		Restricted Access/Cordons	✔
Uneven/Slippery surfaces	✔	Safety Officers/Observers	✔
Smoke		Sectorisation	
Stacking/Storage		Sheilding	
Struck by Object	✔	Signage	
Temperature		Specialist Officers	
Unstable Ground		Supervision	
Ventilation		Traffic Management	
Vibration		WAH Equipment	
Violence		Water Curtains	
Water			

Note: Any sub activity which results in a Red (high risk) or Amber (moderate risk) outcome should only be carried out for risk critical tasks.

Analytical Risk Assessment

An analytical risk assessment should be carried out at operational incidents as soon as time or resources permit.

Likelihood x Severity = Risk

Measures of Likelihood (Probability)

Level	Descriptor	Chance	Description
1	Very unlikely	0 to 9%	The event may occur only in exceptional circumstances
2	Unlikely	10 to 29%	The event could occur infrequently
3	Possible	30 to 69%	The event could occur at some time
4	Likely	70 to 89%	The event is expected to occur in most circumstances
5	Certain	90 to 100%	The event will occur in most circumstances

Measures of Severity (consequence)

Level	Descriptor	Description
1	Negligible	Minor cuts/abrasions requiring minimal treatment. Causing minimal work interruption. No financial loss to the service. No environmental consequences
2	Slight	Injury requiring first aid treatment causing interruption of work for 3 days or less. Slight financial loss to the service. Slight environmental consequence.
3	Moderate	4–14 day lost-time injury(s). Medical treatment required. Moderate environmental implications. Considerate financial loss. Substantial work interruption
4	Major	Major injuries, including permanent disabling injuries of over 14 days. Severe environmental implications. Serious financial loss. Major work interruption
5	Catastrophic	Single or multiple deaths involving any persons. Devastating environmental implications. Huge financial loss. Disastrous work interruption

Risk Assessment Matrix – Level of Risk

Catastrophic	5	10	15	20	25
Major	4	8	12	16	20
Moderate	3	6	9	12	16
Slight	2	4	6	8	10
Negligible	1	2	3	4	5
↑ Severity	Very Unlikely	Unlikely	Possible	Likely	Certain
	Likelihood ⟶				

Minimal	Tolerable	Moderate	High

Risk Groups

A – Operational Personnel, **B** – Support Staff, **C** – Public, **D** – Other Agencies

The hazard consequences and likelihood ratings are multiplied and the resultant number will allow the assessor to grade the risk into one of four categories:

- Minimal (1-3)
- Tolerable (4-6).
- Moderate (8-12.
- High (15-25).

All categories other than that of 'minimal' will require additional control measures to be put in place. Once this has been done, then the ARA must be reviewed and the risk rating should be reduced as a result of additional control measures.

The numerical values can be more extensive with risk and severity gradings of up to 8 each. This provides a more granular level of detail but, fundamentally, the allocation of grading is a subjective task and so more time consuming and can potentially delay reaching the point at which control measures are decided upon.

Risk control hierarchy

The HSE and others have developed a hierarchy of risk control measures that range from the absolute removal to the introduction of PPE and other equipment. These measures, in order of reducing effectiveness (due to increased levels of human input), are:

1. Elimination – redesign the job or substitute a substance so that the hazard is removed or eliminated.

2. Substitution – replace the material or process with a less hazardous one.

3. Engineering controls – for example, use work equipment or other measures to prevent falls where you cannot avoid working at height, install or use additional machinery to control risks from dust or fumes or separate the hazard from operators by methods such as enclosing or guarding dangerous items of machinery/equipment. Give priority to measures that protect collectively over individual measures.

4. Administrative controls – these are all about identifying and implementing the procedures you need to work safely. For example: reducing the time workers are exposed to hazards (e.g. by job rotation); prohibiting the use of mobile phones in hazardous areas; increasing safety signage; and performing risk assessments.

5. Personal protective clothes and equipment – only after all the previous measures have been tried and found ineffective in controlling risks to a reasonably practicable level must personal protective equipment (PPE) be used. For example, where you cannot eliminate the risk of a fall, use work equipment or other measures to minimise the distance and consequences of a fall should one

occur. If chosen, PPE should be selected and fitted by the person who uses it. Workers must be trained in the function and limitation of each item of PPE.

There is a further measure that has been adopted by some, and that is the use of discipline to enforce the compliance with risk safety rules. This has included the immediate dismissal of someone found to be breaking safety rules in high-risk industries. While it is possible to use this approach in the FRS, it is unlikely to be adopted as there are other measures in place to ensure incident ground health and safety.

Completion of the ARA

Responsibility for ensuring that the ARA is completed rests with the IC. At small incidents it may be delegated to a supervisory commander or firefighter who has been properly trained and qualified, or under the supervision of a qualified officer. At larger incidents, where a safety sector has been established, the responsibility for completion will rest with the safety sector commander (SSC), who in turn is responsible to the IC. In addition, the SSC will manage the ARAs from each sector and compile them into an overall incident ARA. It is important that this document is completed by a competent firefighter/officer and regularly reviewed. The time of each review should be noted on the assessment itself, which can be used as contemporaneous note of risk on the fire ground and the measures implemented to reduce that risk. The review should take place between every 20 and 30 minutes, and whenever conditions change on the incident ground that may raise or lower the risk.

For incidents of an almost static nature, such as a controlled burn, a barn fire or standing by to carry out operations at a later time, the IC may choose to extend the period between reviews. The use of a timeline that indicates when ARAs are due for review greatly assists ensuring that ARAs do not become out of date and redundant.

It is essential that significant findings of an ARA are broadcast to command support and transmitted across the incident ground, particularly to operational sectors that are operating within the hazard area. Other working areas where changing risks may have a significant impact on their ability to complete their assigned tasks (such as a water sector at a wildfire whose activities may have to be changed due to a change in wind direction) should also be advised.

All completed versions of ARAs (both sector ARAs and the ARA used for the overall incident) should be retained for debrief purposes as they hold vital operational risk information which will help understand the dynamics of the incident and can be used for supporting any operational learning that is derived from the event.

Advantages of an analytical risk assessment

The advantages of carrying out a written ARA is that it is comprehensive and systematic, taking account of many risks and hazards that may have been missed out the more active and dynamic phase of an incident. Consequently, all details are recorded, which means there is a permanent record of the incident, the nature of the risks and how it changed during the evolution of the incident.

Many services have developed sophisticated processes and policies to assist in the management of risk through ARAs. These include guidance that can be used on the incident ground to assist those completing the forms to derive risk ratings. Because the ARA is a process that requires consideration, it allows those completing the forms time for reflection on the risks and also enables them to discuss issues with their peers and the SSC. In this way a more robust assessment of risks is provided and, subject to identified risk and hazards being communicated to the incident ground, a safer workplace is ensured. Because the process is reviewed, rather than started from scratch each time, subsequent ARAs should be quicker to complete, although it is important that the process isn't treated as a 'tick box exercise' as it is very unlikely that hazards and risks in a dynamic environment – fire, road traffic collision, hazmat incidents – will remain static.

At mulit-agency incidents, the ARA findings should be shared with all agencies and organisations in attendance as this will help them have an appreciation of the risks facing the fire and rescue service and may also prompt them to take specific actions in order to support their own staff at the incident.

Potential problems with an ARA

As with all risk assessments, whether qualitative or quantitative, the degree of subjectivity on the part of the assessor will invariably affect the outcome and the identification of additional control measures. Subjectivity will remain an issue and it is important that the person carrying out the risk assessment has been fully trained in the ARA process and fully appreciates the nature of risks and hazards that may be found at an incident. This individual must also be up to date with risk assessment skill training and experienced in incident operations. As the process is more discontinuous than the dynamic risk assessment (and also more complex) it is important that individuals carrying out the ARA remain focused on that task and are not given other activities while undertaking this role. Allocating the risk assessment to an unqualified or inexperienced individual may seriously compromise the assessment process and increase the risks for those operating in hazard areas on the incident ground. Wherever possible, the person carrying out the ARA should be appropriately qualified by an accredited third-party to ensure consistency of approach across the service and matching industrial benchmarks.

Personal or individual risk assessments (PRAs)

Both dynamic and analytical risk assessments are carried out as part of a structured risk management system to protect firefighters on the incident ground. For many operational activities, however, firefighters will be beyond sight and the direct supervision or control of commanders. Such circumstances include undertaking breathing apparatus operations within a building, or when additional PPE measures such as gas-tight suits are being used, gaining entry into confined spaces, and when the geographical spread of an incident means that firefighters or small groups of firefighters are working in isolation. The final layer of risk assessment focuses on the risks as assessed by the individual.

> *'Firefighters will encounter unexpected or unforeseen situations. Personal (or individual) Risk Assessment is the process undertaken to identify hazards and determine the level of risk they will accept. The outcome of an individual risk assessment will inform and influence their decisions.'*
> HSE

There are a number of models that can be used by firefighters, including the STAR model (below), which provides a systematic approach to an encounter with unforeseen or unexpected life-threatening circumstances where a decision is required. The STAR individual risk assessment/individual decision-making model comprises four steps.

Box 10.1: the STAR model of personal risk assessment

STOP

Stop any intended actions as soon as the new situation becomes apparent. Focus on the new circumstances.

THINK

Think about this new situation and assess the surroundings.
Is there a solution?
Is it clear what to do to reduce the risk of injury or harm?
How can you and those you are with stay as safe as possible in the circumstances?

ACT

Once aware of the situation now is the time to **ACT**.

If it is not clear what the safest course of action is, then the best thing to do is to communicate and get help and warn others that you are not adequately equipped, or that it's not within your range of skill or experience.

Personal risk assessments should fill a void in the risk assessment strategy. Taking a structured approach to personal risk assessment and personal safety has several advantages, including the fact that it is a personal assessment of risk and decisions will be based on a number of factors not least of which are individual experience, training and personal appetite for risk. Proper consideration of risks will inform an individual how to react to the new situation and force them to consider the benefits of taking a course of action, balanced with the additional risk that they face by taking that action.

The relative weakness of a personal risk assessment is that it is unlikely to be recorded and may be missed out or forgotten during a debrief and key risk information may not be passed to those carrying out dynamic or analytical risk assessments.

This risk acceptance will vary from individual to individual depending on factors such as understanding of the situation, the hazards posed, and the individual's knowledge and experience of dealing with similar situations. Individuals will have varying levels of understanding of the hazards and risks that they are facing. When faced with new or unfamiliar situations, a relatively new firefighter may take a different course of action to a long serving veteran who has previously dealt with a similar situation. This may have implications for deployment of crews, and watch commanders must ensure that there is a balance of experience for teams, particularly when crews may be working beyond a reasonable level of supervision due to operational conditions and circumstances.

It is also possible that less experienced staff may be relying on individuals with more experience to provide interpretation, which can result in misinterpretation of the situation. Finally, the appetite for risk can vary depending on the individuals involved. An individual that has a high appetite for risk may decide that the possible benefits of a course of action greatly outweigh the risks associated with that course. At the other extreme, a firefighter with a low appetite for risk may decide that the risks are not outweighed by the benefits and that a different course of action should be sought.

The implications of personal potential risk are very significant and can affect team and sector activity. Indeed, the whole tempo and approach to managing an incident can be determined by the level of risk which individuals, supervisors and even the IC will accept.

Personal risk appetite

Each individual will have his or her own attitude to risk. These attitudes exist on a spectrum and depend on the amount of risk that an individual will tolerate. There are several categories used describing individual attitudes to risk. These are classified as 'risk paranoid', 'risk averse', 'risk tolerant', 'risk seeking', and 'risk addicted'. An individual's attitude to risk can manifest itself in personal behaviours that may influence others and have consequences on the incident ground. Attitudes vary from person to person and depend upon a number of factors including the individual themselves, the situation and the organisation for which they work.

Individual factors

Individual factors comprise the knowledge, skills and attitude that are inherent to the individual in question. These include individual risk tolerance level and the degree to which they are accepting of the risk of harm to themselves and possibly to others. Some firefighters may enjoy the thrill of entering a building on fire and actively seeking to have a visceral experience at every incident. Conversely, a firefighter who takes a more cautious approach to incidents may be more circumspect approaching fires and other incidents to the point where intervention is delayed or not attempted at all.

Other individual factors that affect risk appetite include:

- Their communication and investigatory skills to gather, understand and disseminate information – an individual's perception and understanding of situations and the ability to project consequences of given actions on an incident.

- The ability to interpret and process information within a pressurised environment or situation, particularly where life and safety is at risk. The ability to remain calm and dispassionately assess risks and their potential impact is a key skill in ensuring that correct actions are taken at the right time.

- The more an individual understands incident ground health and safety (knowledge and understanding of risk assessment; safe systems of work; hazards and associated risks) the more comprehensive and wider application of risk management processes they will be able to apply, whether at a personal, team or incident level.

■ A good, professional firefighter's knowledge of standard operating procedures, operational tactics and tactical controls that can implemented to mitigate or reduce risks is essential in understanding and managing risk levels and allowing appropriate measures to address the task in hand in a safe manner. Firefighters with a lesser understanding of these procedures, or less experience due to less time served (or the descent into unconscious incompetence), may have fewer options of firefighting tactics or control measures to consider. As a result, it is possible that there may be a blindness to risks that may hinder operations and increase potential harm to firefighters.

■ The current mood of an individual, the level of stress they are under at an incident, or sometimes external personal stressors such as childcare stress, and external influences, fatigue and the individual's capacity for resilience, all contribute to the individual's risk appetite and decision-making capability at any particular moment.

Situational factors

Situational factors refer to the impact on individuals of the incident that she or he is dealing with at any particular time. As people gain more experience of using incident command tools such as the joint decision-making model and decision control processes, they will inevitably become more comfortable with them and become more confident in making decisions based on their use. Individuals can build on previous experience to tackle new incidents with confidence and make adjustments to deal new situations as necessary.

Depending upon the speed of developments at an incident, the use of a dynamic risk assessment is a quick solution to help identify risks and select tactical mode options, or a more analytical approach can be taken, where time is available. Depending on which approach is taken, this can also have an impact on the risk appetite of an individual at a particular point in time. Where the IC feels that the dynamics of the incident are such that there is an elevated risk to firefighters it may be necessary to call a halt to current operations, regroup and undertake a more detailed ARA before subsequent deployments occur.

The time of day and weather conditions, the availability of resources and delays in attendance will increase pressure on an IC to adopt a more defensive approach as risks may be increased as a result.

Other external factors, such as a large number of members of the public present, including stressed occupiers, may forcibly increase the tolerance of the risks being taken due to the moral pressure on the IC and the potential for casualties. Antagonising factors such as abusive or aggressive members of the public or occupiers may also have an impact on the IC's appetite for risk and may result in a change in risk attitude (either increasing or decreasing appetite), and may cause stress on the IC who will be forced to accept risks that are outside her/his comfort zone.

Organisational factors

The risk management culture of an organisation is key to ensuring a balance between risk and benefit is maintained at all incident types. A safe firefighter should be a more effective firefighter, and where a positive safety culture permeates the organisation, staff are more confident in taking appropriate risks because they have confidence in the culture and training of the organisation. Where investment in safety and training is weak, there is sometimes a collective failure in confidence in the capability of the organisation to manage an incident ground safely, resulting in a risk-averse approach to incident management.

Box 10.2: A case of risk aversion

In 2009, a fire occurred in a large detached house in a rural location in the midlands late at night. The owner, hearing a noise in his attic in the early hours of the morning, investigated the source reach. On raising the access flap to the attic he saw some smoke and evacuated his home, calling the FRS soon after. When the first appliances attended, he was asked by the IC if there was anyone in the house and upon being told everyone had evacuated began assessing the incident, which by now had deteriorated. The owner told them how to get into the building and the attic which was still accessible. At this point the IC decided (on the grounds that there was no risk to life and that entry into the building put his firefighters at too great a risk at that time) to fight the fire from outside and request additional resources. By this stage, the owner was very agitated and threatened to enter the premises at which point he was advised that it was not safe to do so and that the IC would exercise his powers under the FRSA to stop him entering the building. The fire grew rapidly and destroyed the building. The owner sued the FRS for negligence and was awarded a substantial sum to rebuild the home and loss of property. It is believed that the IC misinterpreted guidance that stated that 'firefighters will risk their lives a little to save saveable property' and decided that any internal firefighting would be too risky.

It was determined that had an early entry made into the building to attack the fire, the property would have been saved.

Lessons learned

All firefighting activity is inherently dangerous but some risk must be tolerated. An early attack on a fire may be more dangerous initially but less dangerous in the long run.

The latest guidance takes a more nuanced approach and makes it clear that the balance between risk and benefits is not straightforward. At this incident, the IC avoided taking a calculated risk. This was an understandable decision given the context at the time – four firefighters had died in a warehouse fire in Warwickshire and three officers (who commanded the incident) had been arrested and detained by police for their part in the fire – and fought the fire defensively.

The culture of an organisation will therefore have a major influence upon the individual and their attitude to operational risk. Following incidents where major injuries have occurred, there may be a period of 'nervousness' about dealing with certain types of incident. Such events include unexpected explosions of compressed gas cylinders, which can lead to extensive adoption of defensive modes of operation that may be totally unnecessary under the circumstances. Problems associated with the use of breathing apparatus guidelines may influence decisions not to use them, even when the benefit they may bring is important.

The converse is equally true, however: stung by criticism for being overly defensive or 'risk-averse' following critical incidents, it may be the case that a fire and rescue service attempts to push the level of risk acceptance. A fire and rescue service with a strong safety culture and leadership will be in a better position to moderate shifts in risk taking and ensure that a balanced view is taken and extreme positions are avoided.

In a training environment, firefighters and ICs will be given the opportunity to learn and practice the skills they need to perform effectively on a regular basis. Familiarity with tactics, procedures and operational risks will increase the confidence of a commander when difficult decisions are required. A positive training culture will support the development of individuals and improve the capacity and capability of decision takers.

The impact of differing levels of risk appetite

The different levels of risk that an individual will accept – whether a firefighter, supervisor, tactical commander or strategic commander – will have an impact upon how incidents are managed and how the balance between risk and benefit is perceived. Most people take a risk-tolerant approach, but there will be individuals at the extremes of the bell curve and it is these extremes that can have a severe impact on the running of an operation. Summarised below are the main categories of risk appetite as identified by Hillson and Murray-Webster.

Risk paranoid

This is one extreme of the spectrum, where risk is seen all around when there is in fact no risk at all, or risks are dramatically exaggerated: an IC expressing risk paranoia will have a significant impact on the way operations are run. Some of the outcomes of this extreme behaviour being exhibited by someone in command may include:

- No or limited offensive actions being considered.

- Hazards and risks not being identified or correctly assessed.

- Delays in committing crews or starting offensive actions.

- No instigation of safe systems of work and tactical controls.

- No decision making – uncertainty about which course of action to take.

- Deterioration of the situation due to lack of commencement of operational activity.

- Limited or total loss of visible and assertive leadership being displayed and no command and control protocols put in place.

- IC being ignored and overridden by less senior staff leading to a degree of fireground freelancing.

- Incident escalation making eventual offensive tactics harder to be implemented and achieved.

- Casualty survival may be put in doubt because of a lack of offensive operations, survivability potential decreases.

- Potential for litigation and reputational damage.

Risk averse

Firefighters and commanders who are risk averse have a limited acceptance of risks, but will tend towards caution. While not as extreme as the risk paranoid, there will be a reluctance to take action and the inherent delays caused by this will result in additional damage to property and potentially escalating risk to life, for members of the public and for firefighters due to worsening conditions at the incident. Risk aversion has become a term of abuse within fire and rescue circles and within the tabloid media following incidents at which firefighters have not carried out rescues due to policies and protocols forbidding intervention until suitable safe systems of work can be implemented, such as the attendance of specialist appliances and trained crews.

Some of the characteristics exhibited by those who are risk averse either as firefighters or, especially, those in command include:

- Some offensive actions are taken, albeit reluctantly.

- There are delays in preventing escalation of the incident leading to a worsening situation that increases risks to firefighters and the public alike.

- Frustration in the delay implementing a tactical plan may potentially result in 'freelancing', with other, more junior commanders taking independent actions, or with individual firefighters acting as they see fit in an absence of command leadership.

- Trust in the IC as a leader or manager is questioned, which can lead to subsequent failures to obey commands at future incidents or even during non-operational situations such as at stations.

It is worth noting, however, that some risk-averse operational activities instigated while awaiting the arrival of supporting resources can have a positive impact in terms of incident management. This includes activities such as setting up water supplies, preparatory measures for offensive tactical measures including a command structure, cordons, casualty handling areas, sectorising the incident, and, of course, information gathering. It is therefore important to recognise the difference between *appearing* to be risk averse and *actually* being risk averse.

Risk tolerant

Individuals who are risk tolerant are generally capable of determining the balance of risks and likely benefits in most situations. There is a good understanding of the risk assessment processes, both dynamic and analytical, with decisions made in compliance with the firefighter maxim. Firefighters and ICs who take a balanced view of risk tend to make sure that positive actions are instigated and incidents managed in a progressive manner, with safe systems of work being identified, implemented and reviewed. There is generally a high level of tactical control, and command is suitably exercised across the incident ground. A clear plan, which is communicated effectively and understood at all levels of command, should mean working relationships will be effective and a trust developed between the IC, the supporting commanders and their teams. The level of control and assertive leadership presence should ensure that no freelancing takes place and that the incident plan is being followed, monitored and reviewed on a regular basis.

Risk seeking

Individuals who are risk-seeking may expose themselves, other firefighters, or their teams to excessive risks during operational incidents. There are a number of ways in which this attitude may impact on others at incidents and progress against the plan, and eventually the outcome of the incident.

Some of the possible consequences of having someone with a propensity for risk seeking among your team, or when an IC has a risk-seeking disposition, include the possibility of being distracted from the tasks and roles needed to be undertaken to resolve the incident as quickly as possible. This might be as simple as a firefighter being tasked with what he or she may perceive as a 'mundane' task, such as running out a hose or managing a pump at an open water source, but instead deciding to undertake a different, unapproved action in a more 'exciting part' of the incident ground. This freelancing can not only create a situation where an essential (though unexciting) task may not be completed, but also exposes an individual to risk as their last known location is now incorrect and any search, should it become necessary, would be a waste of resources and may delay locating that individual.

Where there is a propensity for risk-taking, there may be an acceptance of working with hazards that an individual with a balanced approach to risk would deem unsafe. Visible demonstrations of risky behaviour on the part of the IC can encourage others to adopt similar behaviours. This disregard for higher risks may encourage others to take actions and options which expose them unnecessarily to danger and can permeate, not only on the incident ground but within the service itself, creating a risk-taking culture that could lead to a rise in the number of accidents and injuries across the service.

Risk addicted

As with many professions, particularly those that are exposed to hazards on a regular basis, there may be individuals who could be considered 'risk addicted'. These are individuals who actively seek threatening situations that provide an extreme challenge and 'an adrenaline rush', and an opportunity to 'prove themselves' in challenging environments. Individuals who exhibit such behaviours threaten the safety of themselves and others by exposing themselves to unacceptable and unnecessary risks. Not only does this risk the lives of firefighters, members of the public and other emergency services, it also risks reputational damage to the service itself, and potentially exposes the organisation to litigation.

Managing extremes of risk perception

Inappropriate attitudes towards risk can have serious consequences both on the incident ground and in day-to-day work activities. Strong and assertive leadership across the incident ground will help support effective incident management and will limit the impact that incorrect risk attitudes can have. Techniques to manage and control risk attitudes include:

- effective communication of the IC's intentions and plan down the chain of command

- tight monitoring and supervision of crews

- additional monitoring of areas that are particularly hazardous by safety observers tasked with specific areas of concern

- developing an incident ground culture that balances risks with the likely benefits and ensuring that subordinate commanders are clear as to the IC's intentions

- the early identification of hazard areas and specific risks

- deployment of safety observers, safety managers and safety sector commanders to ensure that the priority for firefighter safety is visibly known at all times

- corrective measures for unsafe practices should be endorsed by the IC and all subordinate commanders

- where unsafe practices are identified, it may be necessary to make note of those with a view to taking further action upon completion of the incident, which may include additional training.

Where there is a problem with risk aversion, it may be necessary to encourage those unprepared to undertake hazardous activities and challenge information and negative assumptions. The IC should attempt to explain the plan formulated, any safety issues and control measures already established as part of their and others' situational awareness and risk assessments. They may be assisted by the safety officer or safety sector commander. Sometimes developing a more robust plan using others' ideas may gain confidence and support from those less confident of exposure to risk.

However, as operational activities often take place in a time limited environment, discussing options with a reluctant and risk-averse subordinate cannot continue indefinitely and at some point the IC will need to bite the bullet and use other staff to carry out such activities. Where this is the case, individuals failing to undertake a task should be removed to another area. They should be given another task to avoid giving an unreasonable rise in concern in others, which could affect confidence in the IC as a leader, and in their plan. Any issues emerging from such a situation should be addressed by such health and safety training and support as may be necessary, and other measures depending on the event, such as retraining or a disciplinary investigation.

Summary

You should now understand the following:

The 'safe person' concept and the requirement for effective supervision on the incident ground.

The requirement for various types of risk assessment including the dynamic risk assessment, the analytical risk assessment and the personal risk assessment (including the STAR model of personal risk assessment) on the incident ground along with an appreciation that they should be completed in accordance with the needs of the incident.

The processes of analytical risk assessment and the need for it to be completed by a competent person during an incident. This person is likely to be an appropriately trained manager with appropriate qualifications and experience.

The risk control hierarchy and how reviews on the incident ground reduce risk to a tolerable level.

Appreciate the advantages and limitations of using various types of risk assessment available on the incident ground.

The personal risk appetite of individuals and how that may impact on events during incidents. This includes individuals who have an affinity for risk and those who have risk aversion, and methods of managing those individuals.

Chapter 11: Safety on the incident ground (part 2):

Striking balances, heroic actions and operational discretion

Introduction

There have been incidents in the recent past, unfortunately involving the serious injury or even deaths of firefighters, that have led to the prosecution of fire authorities, individual firefighters and ICs, under both health and safety and criminal legislation. While lessons have been learned and improvements made, there have been concerns about the rationale for investigations, and the processes and the evidence which is used to support prosecution. In order to help clarify the situation regarding the application of the health and safety law in the operational environment, the UK Health and Safety Executive published several documents to support FRSs in understanding the procedures and circumstances under which investigations will take place. This chapter will examine these documents in some detail as they provide guidance about how the HSE assesses the approach and actions undertaken by ICs and others in managing incidents. It will also raise some important considerations about the impact of individual acts of heroism under extreme circumstances. Finally, the issue of 'operational discretion' – a relatively new concept in name but not in fact – will be examined to identify the consequences of such a policy on collective acts beyond operational policies and procedures on the incident ground.

Firefighter safety maxim

In order to provide firefighters with a single guideline that applies to all operational circumstances, the UK fire and rescue service has derived a set of principles that gives those taking part in incidents a good understanding of how to strike a balance between risks and benefits that is proportional to the nature of the incident and its outcomes. This guideline (or tenet) is known as the firefighter safety maxim. It states:

> 'At every incident the greater the potential benefit of fire and rescue actions, the greater the risk that is accepted by commanders and firefighters. Activities that present a high risk to safety are limited to those that have the potential to save life or to prevent rapid and significant escalation of the incident.'
> National Operational Guidance (2015)

In essence, the maxim, while acknowledging that firefighting is a dangerous occupation, clarifies that where life is at risk and there is a chance of that life being saved, firefighters will be expected to risk their own lives. It anticipates that the risk taken will be proportionate to the likelihood of success of that operation. It is impossible to come up with a defined, quantitative balance between the two aspects, but the simple rule is that the greater the possible benefits, the higher the level of risk that a firefighter should take in order to achieve that benefit. For example, a fire in the ground floor of a terraced home with people trapped upstairs and requiring rescue may justify the exposure of two or more firefighters in order to attempt a rescue using breathing apparatus and a hose reel jet. In a more extensive fire in the same building that has involved the bedrooms where people were sleeping, and where there is a strong likelihood of structural collapse, the likely benefits will be greatly reduced and the likelihood of harm to firefighters attempting a rescue will be greatly increased. In this case, it may be that the decision not to send rescue teams in without additional protective measures may be justified.

As the incident risk changes, the outcomes of the DRA will be modified and the balance of risk and benefits will continually change. The IC will constantly be evaluating the balance, and the operational tactics will similarly change to take advantage of a reduced risks to fire fighters (e.g. the ventilation of a compartment fire allows access to a bedroom) or an increase in the potential benefits, such as the discovery of a casualty within a building that was thought to be empty.

The maxim forms a key part of the decision control process (DCP) in that through constant review – 'active monitoring' – the risks versus benefits are continually evaluated, which enables the IC to justify operations which may be more or less risky.

Striking the balance

Following a number of serious incidents that resulted in the deaths of firefighters, there had been concerns about the enforcement actions taken by the HSE and police service, including the prosecution of individual fire officers and fire authorities. This concern led to a clarification of the situation and resulted in the publication of the documents *Striking the Balance Between Operational and Health and Safety Duties in The Fire and Rescue Service.*

This document established the principles by which an investigation will be carried out by the HSE, enabling fire and rescue employers to comply with health and safety requirements and protect the public and their own staff during emergency operations. Among other things, it aimed to:

■ help firefighters and managers understand the practical application of health and safety law with respect to the fire service's operational work

■ ensure consistency of approach by the HSE

■ promote a safety culture that properly balances risk and proportionate measures to resolve incidents

■ set out expectations on fire and rescue authorities as employers

■ provide a mechanism to support learning from incidents and developments.

The HSE recognised that the FRS operates in difficult environments into which they have sent firefighters to save lives and protect property. It also recognised that members of the public have an unrealistic expectation of the risks to which firefighters will go, despite having little hope of accruing any benefits. Because of the uncertainty of fire dynamics (and those of other incidents), firefighters may sometimes be confronted by situations outside their actual experiences and have to make decisions in 'fast moving, pressurised and emotionally charged situations, not of their making'. They may often be operating in environments where complete information is not available and their ability to control all aspects of their workplace is limited.

The reason that this is important is that within the fire and rescue service there had been a fear that prosecution for health and safety failures was likely to become common, and that there was a pervasive misunderstanding about the risks faced on incident grounds which meant that ICs, subordinate officers and fire authorities would be prosecuted unfairly as a result of accidents and injuries. *Striking the Balance* was an attempt to reassure the FRS and staff that these fears were ungrounded by recognising that, even when all reasonably practical measures to mitigate and minimise risk are applied, there is still the possibility that harm may occur.

It is therefore assumed that, as far as is possible, FRAs deliver an effective service, underpinned by extensive risk assessments and measure health and safety performance together with a process for learning from operational incidents to prevent future accidents.

The Health and Safety Executive works with fire and rescue services to promote health and safety, and to improve HSE inspector's knowledge of the service. Importantly, when investigating incidents, the health and safety executive inspectors will consider the following:

- The adequacy of plans and policies, risk assessments and procedures employed by services in order to deal with foreseeable incidents.

- The adequacy or level of reasonableness of the fire authority's actions under the circumstances.

- The appropriateness and effectiveness of crime control systems in place and during operations.

- The contributory nature of lack of preparedness to deal with a foreseeable risk in any particular circumstances.

- The actual information that was available to firefighters at the time of an accident. This is important, because it was felt that inspectors (and others) reviewing incidents in the light of information that became available after the event was unfair – reasonable decisions based on the information available at the time (including generic risk assessments, site-specific risk assessments, dynamic risk assessment, analytical risk assessments) will not be picked over in the light of information that would not be available until after the incident. *The HSE therefore states that 'inspectors will not revisit decisions made during operations with the benefit of information that could not reasonably be known at the time'.*

- The preparation that individuals had to make quality decisions prior to the incident.

- The degree to which arrangements made to fulfil the statutory duties of the service were sensible, effective and practical. It also looks at the risks taken in order to deliver the outcome and whether they were reasonable under the circumstances.

- How fire and rescue authorities prepare ICs and firefighters for operations through information instruction and training.

The net effect of this document was to give a balanced view of what the HSE would be looking for in the event of an accident occurring. It did not mean that prosecutions are unlikely to take place in the future, but it did give a proportionate response to events. The HSE will not seek to 'second-guess' decisions made in pressurised circumstances when situational awareness was not as complete as it would become later in an incident. For FRAs, the HSE document recognises that, to enable an adequate response to the management of incidents, a great deal of pre-incident work needs to be completed.

This document should not be taken as a carte blanche to take unreasonable action in any circumstances, but it should remove some of the concerns of incident commanders at all levels about how decisions they make will be viewed after the event.

Heroism in the FRS

Following a number of reports criticising the inaction of firefighters at incidents where members of the public died, it was felt that guidance should be provided to enable 'heroic acts' to be addressed to reduce the risk of firefighters being subject to investigation or prosecution under health and safety laws.

While the statutory duties on employers to manage all foreseeable risks effectively and review incidents in order to learn from those experiences remain in place, and employees (firefighters) are expected to act 'sensibly and responsibly' and should not act 'recklessly', it is recognised that due to the nature of operational incidents, firefighters often undertake great risks without receiving specific orders and there are no relevant civil work procedures for them to follow. While such acts of heroism should not be recognised as being behaviour that is expected by employers on a regular basis, they are rightly rewarded by fire and rescue services.

The criteria the HSE has set to view the actions of a firefighter as being heroic are:

- The firefighter(s) have decided to act entirely of their own volition.

- They have put themselves at risk to protect the public or colleagues.

- The individual's actions were not likely to put other officers or members of the public at serious risk.

Upon notification of an incident in which a firefighter has been seriously injured, the HSE will make initial enquiries and carry out an investigation into the FRA's operational arrangements and how it manages health and safety. Where the incident involves an act of heroism, which may be partially or wholly responsible for the accident, then the individuals will not face an investigation into their actions. Furthermore, where an act of heroism results in injury or death to someone other than the firefighter acting heroically, the fire and rescue authority will not necessarily be prosecuted as the individual was working outside the service's guidance and policies.

The issue may be slightly complicated if the heroic act was so far outside of normal practices and procedures that the individual has, in fact, carried out a 'frolic of his/her own volition'. Where the actual act has been followed to achieve the FRS' aims (i.e. the rescue of someone or to stop escalation of a fire), vicarious liability for the individual applies. Where a firefighter has been injured undertaking an action that goes beyond the boundaries of reasonable risk-taking (i.e. reckless) the vicarious liability for the individual's actions may be limited.

Due to the limited number of cases in this area it is unlikely that a definitive answer as to whether any specific act is deemed heroic, foolish or a frolic will be provided. What this guidance should not do is to encourage individuals to undertake reckless activity on the incident ground in the name of heroism – and it is up to the IC and other commanders to ensure that, wherever possible, adherence to existing standard operating procedures and guidance should be the norm.

To date, no FRS or individual has been prosecuted involving an heroic act, and the loss of firefighters in such circumstances, attempting the rescue of members of the public, has quite rightly been treated as a death on duty in heroic circumstances.

Some examples

Example 1
Two appliances with a combined total of eight firefighters attend a house fire where persons are reported missing, last believed to be in the front bedroom. While one crew are pitching a ladder to the bedroom window, a breathing apparatus team get ready to take a hose reel jet to fight the fire and make their way to the bedroom to hand casualties to the ladder crew. As they are about to enter through the front door they test their hose reel jet and find the water supply has failed. They hear voices screaming from the first floor and instead of waiting for a replacement jet, they enter the property and make their way to the bedroom where they find a casualty and hand him to the ladder crew. They make their escape down the ladder.

In this example, the firefighters were able to identify the location of the casualty based upon prior knowledge and on the screams from the first floor. When the water failed, there may have been a delay in entering the building with consequential additional distress and potential injury for the casualty. The IC had not directed them to enter the building without water: rather, they had acted on their own volition and accepted an additional risk in order to protect a casualty while not putting other firefighters or members of the public at risk. This could be considered to be an heroic act.

Example 2
For this example we use the same scenario, but with the breathing apparatus team consisting of a firefighter and a crew commander (as BA team leader). If, when the water supply failed, the crew commander (team leader) instructed the firefighter to follow him/her into the building to carry out the rescues, then the situation changes.

While the crew commander may be taking an act of his *own* volition, the other team member is being ordered or 'directed' to take action. The crew commander will have carried out a heroic act, but the firefighter will have been following orders and this will not strictly be an act of heroism. In the event of an injury occurring to either, any liability allocated to the organisation will necessarily be greater if the impact was on the firefighter compared with that on the crew commander.

Example 3

Three appliances with 14 firefighters attend a fire in the roof space at a two-storey residential care premises. On arrival of the first appliance, the care assistant advises them of eight casualties on the first floor. A crew of two BA wearers with a jet enter the building under rapid deployment and ascend to the first-floor landing and enter the corridor. The corridor splits left and right and smoke is starting to fill the corridor. Realising that as a team they can only search one corridor at a time, and given that reinforcing appliances will not be arriving for several minutes, they have a brief discussion and split the team, with one firefighter going down the left corridor and one down the right. They rescue several elderly casualties individually and take them to the entry control point to receive first-aid by paramedics. The firefighters then return to their firefighting duties within the building.

In this case the firefighters had disregarded one of the basic tenets of breathing apparatus operations – they had split up the team in a building that was on fire. However, they had received no instructions to do this from the IC and they were fully aware of the additional risk at which they were putting themselves. Due to the speed of the operations the rescues had been completed by the time reinforcing appliances arrived and so had not put colleagues or members of the public at risk. It can be concluded that this is clearly a heroic action as defined by the HSE.

Operational discretion

For many years firefighters have been using equipment and 'flexing' operational procedures to meet the requirements of carrying out rescues at incidents. The reluctance to work outside standard operational procedures following a number of high-profile incidents where firefighters and members of the public lost their lives or were badly injured, led to the perception that there was a culture of risk aversion beginning to emerge among many junior and senior incident commanders. Fed by negative reports within print and social media, many of these incidents involved activities and procedures associated with specialist rescues – particularly underground and water rescues. There have been several incidents which have raised the issue of what to do in circumstances that are beyond the experience of most firefighters and outside the 'normal' guidance.

Incident 1: Casualty down a disused mine

In July 2008, while walking home across open ground in the dark, a woman fell approximately 15 metres down a disused mine shaft. Following the arrival of fire crews some two and a half hours later (15 minutes after the first call), the casualty was located and the scene of operations cleared of debris and vegetation and then one firefighter was lowered down the shaft using *safe working at height equipment* to assess the casualty's condition. A specialist heavy rescue vehicle crew supported the planning and supervised the lowering operation. At this point the IC (a watch commander) determined that the balance of risks and benefits justified this action. Having covered the casualty with a space blanket, the firefighter then provided first-aid and oxygen. An attending paramedic was preparing to enter the shaft and was in the process of donning a body harness when a senior officer arrived at the incident.

The initial IC and the new commander discussed the plan and agreed that the rescue would be improvised using existing equipment and secured lines to vehicles acting as anchor points for lowering equipment. During this period, a police sergeant requested the attendance of a police mountain rescue team. A second, more senior fire officer then attended the incident and stopped the attempted rescue on the grounds that he believed (wrongly) that the police mountain rescue teams would be in attendance within 30 to 40 minutes. In fact, the police mountain rescue teams would take around 90 minutes to attend, and following the deployment the casualty was recovered after a further 65 minutes, just over four hours after the first call to the fire and rescue service. Due to hypothermia, she suffered heart failure and died later that morning.

In this case there had been a number of factors that delayed the rescue:

■ There had been guidance issued by Her Majesty's Inspectorate of Fire Services (Scotland) indicating that *'there are circumstances …. where the fire service is strongly advised not to become involved in mines incidents'* and advised that *'normal Fire service would not be suitable'*.

■ An improvement notice had been served on the brigade which stated that on a previous occasion they had failed to ensure a person who was not in their direct employment (members of the ambulance service) were not exposed to risk to their health and safety. In the example above, this may have been the reason for stopping the deployment of the ambulance worker into the shaft.

■ On another previous occasion two firefighters had lost consciousness and required rescue after entering a 13-foot deep hole to rescue a child.

■ Following the introduction of safe work at height equipment, firefighters were instructed not to adapt existing equipment and skills to perform specialist rescues which had previously been commonplace practice.

Incident 2: Casualty in water

In a second incident some years later, a member of the public drowned in a model boating lake after suffering an epileptic seizure. The first attending firefighters refused to enter the water due to its depth, following a risk assessment which led to the conclusion that the casualty was deceased (having being submerged for over 20-30 minutes before their arrival) the IC decided to wait for the appropriate resources to attend. A police officer who volunteered to wade into the lake, which was only two or three feet deep with a concrete bottom and sides, was told by police control not to enter the water because of the opinion of fire and rescue service. Despite the service stating that the decision not to deploy into the lake 'had nothing to do with health and safety or the depth of the water', media comment was highly critical and helped fuel the debate about following service policies and orders rather than deploying using what may be termed as 'common sense'.

These and other incidents – there have been other similar high-profile events ('health and safety gone mad' histrionics in some of the tabloid media) – led to the recognition that the professional judgement of competent and experienced firefighters, termed 'operational discretion', had not been fully utilised at unusual and uncommon incidents or where some aspects of an incident are non-standard.

The rules for the use of operational discretion

Essentially, operational discretion should be applied when adapting standard operational procedures, using the information that is available and using professional judgement when weighing up any additional risks associated with using a modified systems of work. Clearly, this should be undertaken only when the risks and benefits of using existing procedures have been assessed and found to be unfavourable. More formally, operational discretion may apply to rare or exceptional circumstances where *'strictly following an operational procedure would be a barrier to resolving an incident, or where there is no procedure that adequately deals with the incident'*.

There are a number of outcomes for which the deployment of measures under operational discretion are appropriate. These include:

- the saving of human life
- taking decisive action to prevent an incident escalating
- where taking no action may lead others to put themselves in danger.

Operational discretion should only be applied once a risk versus benefit calculation has been carried out by the IC and it has been identified that the additional risk is commensurate with the likely additional benefits that could be achieved.

The IC who is considering implementing discretionary measures to resolve an incident must be sufficiently aware of current operational procedures, and they should not allow anyone under their command to use operational discretion if they don't understand the predetermined protocols. Varying the tactics or procedures because there is a gap in your knowledge is not a justification for operational discretion. A professional officer should know the standard operating policies and procedures before trying to vary them at an incident.

It is important that the IC is not unduly influenced by those personnel under his command to adopt operational discretion. The IC must also be aware of the skills and capabilities of the crew that will be asked to carry out these procedures, including their ability to adapt to modified procedures. It is important to ensure that the operational equipment that is to be used is fit for purpose and not likely to fail as a result of an unintended use.

Where operational discretion is applied it should be used. As short a time as possible, and as soon circumstances allow (achievement of objectives, attendance of sufficient resources etc.), reversion to the current standard operating procedures should take place. It is important that any use of operational discretion be recorded along with the justification for its implementation.

Where an FRS identifies multiple occasions when a similar operational discretion has been applied, then it is likely that the event is not unforeseeable and that a standard operational procedure or policy should be developed to ensure that the operational response is consistent across the service. Identification of multiple circumstances of the same operational discretion being used should be notified to Her Majesty's Inspectorates (in the UK) so they may collate evidence for future national policy development.

It is important to ensure that all staff are aware of the limitations of operational discretion: it should never be used as an excuse for bypassing existing safety precautions and safe systems of work in order to achieve a speedier resolution of the incident while compromising safety. There is also the potential for confusing operational discretion with heroic action: operational discretion is an organisational solution to a challenging incident. Implementing operational discretion means that the organisation is taking a calculated risk by adapting procedures or equipment to achieve an organisational objective following a rigorous assessment. It does not allow firefighters to act on their own volition and so does not constitute a heroic act.

Caveats

It is important that services recognise circumstances in which they may inadvertently be putting firefighters at risk of adopting operational discretion or heroic acts through inappropriate mobilisation policies or though failures to enforce protocols at operational incidents. For example, the increasing use of firefighting vehicles crewed with two or three firefighters runs the risk of them arriving at property fires unsupported by the full attendance of a pumping appliance for some time. Policies that state that offensive interior firefighting is forbidden may not be sufficient to restrain firefighters from undertaking offensive interior operations in circumstances where there is moral pressure from members of the public or where persons are trapped in the building.

It is therefore foreseeable that firefighters may enter premises with less than adequate safe systems of work and put themselves at risk. Under these circumstances, an injury to firefighters it is likely to mean that the service policies are challenged, found inadequate and deemed to provide a less than safe system of work.

There are also circumstances in which it may be unclear as to whether operational discretion or a heroic act has been used. There have been several notable incidents involving large fires where many members of the public have been trapped and the first attending crews have been insufficient to carry out interior breathing apparatus operations. Where BA crews of two have mutually decided to split up after entering the building in order to carry out effective rescues, this is clearly a heroic act. Where they have been *ordered* to enter the building by the IC and carry out the rescues is clearly different: this is the use of operational discretion (subject to an assessment of the additional risks and the likely benefits).

The request for a volunteer to carry out a particularly hazardous action is not a request for someone to carry out a heroic act. The fact that it is a request from a managerial representative (of any level) means that *this is a direction from the organisation* and not an individual carrying out an action of his or her own volition.

These circumstances described highlight the problem of discretionary procedures and heroic acts and the potential for one to be confused with the other. Fire and rescue services and individual ICs at all levels should ensure that staff for whom they are responsible are clear about what constitutes a heroic act and what is operational discretion.

The need for logging and justifying actions taken under operational discretion

According to HSE figures, to date (2018) there have only been a dozen or so recorded cases where ICs have declared 'operational discretion'. Of these, the HSE found that fewer than half were justifiable and that existing procedures could have been implemented to achieve the outcome within the required timeframe. It is therefore imperative that all decisions under the guise of operational discretion must be accurately recorded, with a full description of the rationale together with the reasons why existing procedures were inadequate to deal with the circumstances.

In this way the IC is ensuring that not only are the circumstances recorded, but he is also setting out the case why other alternatives that were considered were not implemented. The more robustly a process is logged, the less likely it is that there will be an error of judgement on the part of the IC and, in the worst circumstances (where a firefighter or member of the public is harmed), then the IC will be able to demonstrate that all due diligence was undertaken before implementing activities.

Summary

You should now have an understanding of the following:

The concept and application of the firefighter safety maxim on the incident ground.

The health and safety executive document *Striking the Balance*, the reason for its creation and the limits of its applicability on the incident ground, along with the implications for ICs and firefighters.

The concept of heroism in the FRS, published by the HSE, and misconceptions of its intent along with the implications of it being used incorrectly.

The concept of 'operational discretion', its utility and limitations particularly when used to improvise operational procedures and techniques outside the parameters laid down by individual fire and rescue services. The importance of logging and justifying actions taken under operational discretion should be recognised as should the limited scope for operational discretion to be undertaken at real incidents.

Chapter 12: Communications

Introduction

Communication systems provide the sinews that hold the incident command system together and help ensure that all operations and tasks are kept co-ordinated at all times. Unfortunately, the failure of communications – both the human and technical aspects – are often blamed for problems that occur during operations: misunderstood orders; misheard instructions; radio signals screened out by metalwork in a building; failures of power sources (by not charging batteries, not having enough batteries etc). The incident commander must have a good understanding of communications at an individual level and at a systems level to make sure that intentions, activities and tasks are all understood and carried out safely.

The importance of communications

Incident communication systems are essential in supporting the incident command framework and are an essential part of all safe systems of work on the fireground. Communications allow the incident commander to:

- acquire the 'real time' information that develops their situational awareness and continually informs the incident management planning process, and maintains the shared understanding at sector and service level (through fire control).

- ensure that activities and operations are co-ordinated and controlled across the incident ground by ensuring that everyone understands the IC's intentions and plans

- share information and situational understanding in order to provide a multi-agency co-ordinated approach to resolving serious incidents (see Chapter 14).

When used in a professional, authoritative and confident manner it can be a tool for fostering and reinforcing a team spirit on the incident ground and provide reassurance for firefighters that the incident is being managed effectively.

Face-to-face communications

Communications between people are usually undertaken at a combination of both verbal and non-verbal levels. The words we say, the tone of our voice and our facial expressions and gestures all add to the receiver's understanding of what we are trying to say or explain. It is therefore important that all commanders consider the gestures and facial expressions they use when briefing staff or when being briefed. Nothing undermines a serious message like a facial gesture that can be misinterpreted by the receiver (such as raising an eyebrow when being briefed, which might be interpreted by those present as a sign that the individual disagrees with the order or task!).

At the incident ground, face-to-face communications may not be possible due to the incident itself, the topography and scale of the incident and the demands placed upon the IC and others. This can mean that communications must be made through other means such as radios and runners, and may need additional considerations if they are to remain effective.

Communicating effectively

The way we communicate with others is key to how a message is received and understood, and how actions are implemented. It is a combination of how we speak, what we say and how we behave when we are communicating. To assist in making sure the communications are effective there are a number of points that should be borne in mind, particularly in face-to-face meetings:

Preparation: before beginning any dialogue, particularly if the message to be given is complex or long, give some thought to what needs to be said and what the outcome is. This preparation may include writing down key phrases, including the use of an aid memoire or briefing tool such as SMEAC (Situation, Mission, Execution, Any questions, Check understanding), or the use of a decision support tool such as the DCP as a briefing template. At larger incidents, the use of an IIMARCH briefing template may assist for briefing other agencies should be considered (see Chapter 15).

Clarity: Messages should be clear and unambiguous. There have been many occasions when both parties have thought they understood an instruction but have interpreted it incorrectly, with unfortunate consequences. Use common terms that are easily understood, particularly if non-FRS personnel are present.

Concise: Communications on the incident ground need to be concise, but not so brief that messages may leave doubt or may be misinterpreted.

Understood: Questioning the receiver's understanding will help ensure that the message has been received accurately.

It is a dialogue: The receiver should be encouraged to ask questions and raise points for clarity and understanding. Even where a small group is being briefed, questions should be encouraged so that a common understanding of the situation is achieved.

It is directive and assertive: It is important that there is no misunderstanding of the purpose of the dialogue. If the discussion has the purpose of identifying a course of action then this should be stated clearly and at the beginning; if it is a tasking meeting, then this should also be stated. Confusion about the intention of the discussion will delay action and cause potential misunderstandings. It must be remembered that the IC is responsible for decisions made and directives issued should be made with authority and confidence.

It should be encouraging and enthusiastic: Incident management is a team activity and the IC who encourages crews to help arrive at a solution will reap the benefits from the combined thinking-power present. Engaging with the teams and encouraging them to develop ideas will help motivate and build team confidence. It is important that the words of the IC is matched by their attitude, demeanour and behaviours, as an inconsistency between either may undermine the message.

Electronic communications

Constantly developing technology now means that electronic communications form a significant means of communications at incidents. Risk information can be downloaded immediately using mobile data terminals and transmitted to laptops or iPads at different sectors; incident geodata can be used to locate pumps hydrants and water sources; and live audio-visual links from aerial or terrestrial cameras can provide live information from all parts of the incident ground. There is also the opportunity to acquire knowledge from other agencies through further electronic transmission.

While there are many advantages to using technology, it is important that the limitations must also be appreciated and that fundamental, alternative methods of incident ground communications are also retained to provide resilience. Some of the problems with high tech solutions include:

■ The resilience of a signal when using radio downlinks cannot always be guaranteed and may be subject to adverse weather conditions (e.g. AUVs may not be able to fly when windspeeds exceed 10km/h).

■ Any item of audio-visual equipment that needs to be left in situ, such as a camera, may be susceptible to theft and it may not always be possible to appoint a firefighter to 'guard' such equipment.

- Remote communications, in particular, are generally dependent upon batteries, which can run down.

- Any computer system is vulnerable to a system crash and may be lost at a critical time.

- The security of a network over which information and data is transmitted must be robust as some material that may be used to manage incidents may be of a sensitive nature.

- Information may sometimes not mean as much to the receiver as it does to the sender – plans that the sender believes are straightforward may be viewed incorrectly by the receiver without additional information.

It is also important to recognise that with these additional means of communications, there is a danger that effective information filtering does not take place and that those in the CSU, sectors and the IC become overloaded with information. There is also the distraction effect of video images in the CSU, which can sometimes lead to a loss of focus of key staff in the CSU as they are drawn to live images of fire etc.

Written communications

It is rare for incident ground communications to be written, but where this is the case messages should be carefully constructed and unambiguous. It may be the case that information about the premises may need to be sent to each sector, and this can be achieved through the use of MDT plans with instructions, diagrams or actions written on them. The use of a CSU's photocopier will assist in keeping a record of actions for future reference and may be referred to in the decision log.

Mobile telephones

The use of mobile telephones is almost universal and they can be an effective tool for communicating at incidents where detailed discussions between individuals is essential. This could, for example, include a conference between hazmat specialists or the briefing of another individual (e.g. an elected member of the local authority) about an incident. Generally, mobile telephones should not be used to pass orders on the incident ground, to send messages to fire control, or used to bypass the accepted methods of incident communication without very good reason.

The use of phones should also be restricted because:

- they may not be secure

- they may lead to misunderstandings as they will be creating a second communications link (the first being the incident ground CSU link) between the incident ground and fire control, which may result in information bypassing commanders and potentially raising the risk level

- key information may not be recorded

- the importance of the message may not be appreciated by fire control as it may be considered less urgent, perhaps, because of the means by which the information was passed – almost conversational

- mobile networks sometimes crash due to usage levels – as may be the case at large, high-profile incidents.

Incident communications

Communicating effectively at incidents is a safety-critical activity, whether requesting additional resources through fire control, checking the air pressure in the BA set of a firefighter in a building or ordering an emergency withdrawal from a building in imminent danger of collapse. Incident grounds are busy places and the communications that are in place to manage the incident need to be clearly set up, identified and recorded, so that the minimum time is taken in transmitting essential information. Each FRS will have its own procedures for arranging networks and radio channels but the principles apply equally for all. Incident communications can be broken down into the communications between the incident ground and fire control and those on the incident ground itself.

Communications between fire control and the IC via Command Point/Unit

The radio link between the incident ground and fire control allows for the dispatch and mobilisation of resources, the strategic deployment of resources to maintain fire and emergency cover and to enable reinforcement of the service, both at incident level and service level through mutual aid and national support mechanisms. Many messages take the form of digital transmissions rather than voice only, but the sequencing and content of the messages remain the same.

Messages to fire control

There is a sequence of messages that should be sent to fire control that should be appropriate for all incidents. Apart from routine messages, the messages are:

First impressions message: This is a short message to fire control giving an early idea of what the incident looks like:

> *'First impressions: house fire, first floor well alight.'*
> Or
> *'First impressions – nothing showing.'*

A clear first impressions message can help oncoming teams begin to consider the likely actions and equipment that may be needed and what tactics might be required at an incident.

Assistance message (also known as a 'make up'): Where the initial resources require reinforcement such as a request for additional pumps, aerial appliances, special appliances or officers, then an assistance message – sometimes sent as a PRIORITY MESSAGE (which supersedes normal radio traffic and alerts control operators) – should be sent urgently.

> *'PRIORITY ASSISTANCE MESSAGE – Make pumps 10, aerial appliance required, hazmat officer required. RVP at Acacia Avenue.'*

An RVP should be given early in an incident to make sure that all responding vehicles arrive at the same location, near but not too close to the incident.

Persons reported: Where the IC finds out on arrival that persons are reported trapped in the building on fire, the priority message 'PERSONS REPORTED' should be sent. This will cause the mobilisation of an ambulance and, sometimes, additional FRS resources automatically.

Informative message: This message should provide sufficient information for fire control and oncoming resources to gain an overview of the incident. The exact format of the message will depend upon the FRS policy but will usually include some or all of the following:

- **Height of building** ('A building of four storeys').
- **Area of building** ('15m x 10m').
- **Use of building** ('Used as an industrial laundry').
- **Location of fire** ('Fire in first floor offices').
- **Equipment in use** ('8 BA, 2 Jets from hydrant, 135 ladder in use, Stage II BA in operation').
- **Tactical mode:** ('Offensive').

Any other information that may be of use to oncoming appliances or commanders may also be given in an informative message.

Types of incident other than fires will have messages that broadly follow the pattern above with a description of the incident followed by a list of equipment in use and tactical mode. This type of message is also used for passing on other details of the incident, including a change of incident commander, a change of tactical mode, a tactical or emergency withdrawal as well as routine information transfers.

When the incident has been contained, an informative message to that effect may be sent. For fire incidents, this may take the form of a 'FIRE SURROUNDED' message; for a hazmat incident this could be 'CHEMICAL LEAK STOPPED AND SPILLAGE CONTAINED'.

Stop message: This message indicates that the incident has been brought under control and no further FRS resources are required (although relief crews may be needed in due course).

> *'Stop for 153 Bath Road, Oxford. Fire in flat, third floor of four-storey building, 20m by 15 metres. 75% of lounge severely damaged by fire and heat, water and smoke damage to flat immediately below; smoke damage to remainder of third floor. Six BA, Stage II, two Jets, thermal imaging camera and PPV fan in use. Tactical mode: Offensive'.*

All messages should originate from the incident commander who should be named at the start of the message.

Incident ground communications

Robust and reliable communications are essential for the safe and effective command and control of the incident ground. Normally, incident ground communications come in the form of handheld radio sets, which provide point to point and incident-wide messaging. Each service will procure their own radio sets and set their own policies for channel use. Most radios are provided with a minimum of six radio channels which may have a pre-designated purpose (such as BA and command channels). A typical allocation of radio channels might be:

- **Channel 1:** Primary or general incident command communications, used initially until the incident grows to the extent that a separate command channel is required.

- **Channel 2:** Incident command channel (to be used when channel 1 is likely to be overused), which requires the attendance of a CSU to set up a repeater/base station with a or 'talk through' facility. Each service will have its own protocols and procedures in organising this facility.

- **Channel 3:** This may be used for BA and other fireground communications but will not always be necessary for smaller incidents.

- **Channel 4:** May be used as a second channel for incident ground operations or dedicated to a single purpose (e.g. logistics, BA support sector, marshalling etc).

- **Channel 5:** Additional command channel, filtered through the CSU.

- **Channel 6:** Additional channel for BA, specific tasks etc.

There are additional radio networks (including inter-agency networks) that may be available in a FRS which can supplement incident ground radios at major incidents.

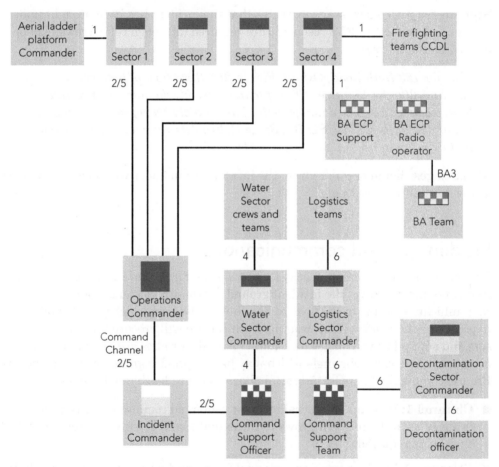

Figure 12.1: Communications at a large Incident

Using hand held radios

Using hand held radios requires the same level of discipline as for main scheme radios on pumps and other units. Correct call signs should be used at all times – 'Sector 1 Commander', 'CE 131 Pump Operator', 'Incident Commander' etc.

Changing radio channels should only be carried out once instructed to do so by the IC or the CSO (or CSU crew on their behalf).

Where communications through radios is difficult, a runner should be used until the problem has been identified and resolved by specialist staff or CSU crew members.

Hand held radios should not be switched on when in the CSU and not used to transmit within 10m of the CSU.

Note: where there is no BA supervisor it may be necessary for the IC to allocate a second radio to the ECO to monitor Channel 1. Alternatively, an additional firefighter may be deployed at the BAECP with a radio (Channel 1) for this purpose or the IC could use the location as the incident command point.

Summary

You should now have an understanding of the following:

The importance of communications both on and off the incident ground including electronic and personal methods.

Incident ground communications including use of main scheme and 'fire ground' radio systems and the use of such systems for sending messages between the incident ground and fire control and that the incident itself.

The sequence of messages from incident ground fire control including first impressions, assistance, informative and 'stop' messages.

Chapter 13: Closing down an Incident

Introduction

Once an incident has been controlled and the intensity of activity and operational concern declines, appliances need to be made available for redeployment, and relief plans are required to replace those crews who have been at the incident the longest time and that welfare measures are put in place. It is important during this phase that the incident commander remains vigilant and ensures that firefighters do not drop their guard as a result of tiredness or from the illusion that the risks are over. The continuing need for safety is a key activity throughout the closing down phase and can only conclude at the end of the incident and upon the return to station. Effective closing down of an incident is as important as any other phase of the operation but it is one that can often be forgotten as things wind down.

Scaling down operations

As the incident becomes controlled, resources required to continue operations will diminish until the point is reached that the incident can be closed. With this reduction of resources, the rank/role of the incident commander can be reduced as well as the number of additional supporting resources at the incident, including safety officers and command support structures.

Reliefs: estimating resource requirements at a declining incident

Identifying the number of relief crews, officers and appliances at an incident can be challenging but there is usually the luxury of time in which to carry out these calculations. At large incidents where a command support team has been in place for some time, they may be the best people to determine the level of resources required to conclude the incident. When identifying the number of pumps, officers, special appliances and firefighters, it is important that fire control are given the details and then left to manage where those resources will be deployed from as they will have a better idea of the strategic availability across the service. They will also be in a better position to manage resources in the context of day-to-day operations and business within the rest of the service.

The calculation for resources will be made on the same basis as decisions made during the growth phase of the incident: identify the tasks and whether the tasks are sequential or if they need to be completed simultaneously, assess resources to complete the tasks, when will they be required at the incident ground what will the incident command structure be now that resources are being reduced. Having completed the requirements, this may be transmitted to fire control for processing.

Relieving crews must be effectively briefed to ensure their safety and operational activity is not compromised. An optimum briefing will consist of a safety briefing by the incident safety officer or team member, a tactical briefing about the overall incident and their role, and then an introduction to the area in which they are to operate. A briefing tool such as SMEAC should be used which will enable individual and team briefings to be focused on their specific tasks and the incident commander's overall intentions. Ideally, relieving crews should be briefed by the sector commanders in whose sector they will be operating and meet up with the crew that they are replacing to ensure that they have a complete understanding of the incident, the geography and risks in the area they will be working in. All three briefings should not be rushed as key information may be missed out or forgotten. Firefighters have been badly injured when they have not been told about key safety aspects as a result of the crews being relieved prematurely leaving the incident ground.

Transferring down command

All transfers of command at a declining incident should be carried out giving the maximum detail possible so that the oncoming commander has as complete as possible an understanding of the scene. While OTHERS (or similar handover mnemonics) may be easy to use at simple incidents, whether for briefing relief crews or handing over command, other information sources should be considered to give a 'richer' information picture to an oncoming commander. This may include:

- the decision log
- the log messages sent
- the analytical risk assessment forms in a chronological area (to give an understanding of the progression of the incident)
- a briefing by the operations commander and command support officer (where available) to ensure completeness

- the SSRI

- a walk-through of future plans, aims and intentions for the incident.

On completion of the briefing, a confirmation of change of command should be notified to fire control and to the incident ground by radio.

Reinspection

One of the most difficult situations at fire incidents occurs when there is limited evidence of burning remaining but the IC wants to be certain (or as certain as can be reasonably expected) that the fire is out. The use of inspections is one method of achieving this level of certainty and usually involves sending a pumping appliance and crew or an officer to inspect the incident over a number of hours to ensure there is no risk of reignition or continuation of burning. The time intervals can be relatively short but may be extended as the likelihood of growth diminishes. Where a 'stop' message has been sent, the commander who carried out the reinspection will send a message to the effect that the 'incident is now closed' which then enables fire control to terminate the incident log. Where a 'stop' message has not been sent, then one will be sent upon closing the incident.

Securing the premises

At the conclusion of an incident it may sometimes be necessary to secure the premises to prevent unauthorised access, theft or further destruction taking place. Under the Occupiers Liability Acts 1957 and 1984, property owners have a legal duty to make their properties as safe as reasonably possible. The fire and rescue service are not responsible for the costs of boarding up whether damage is the result of a criminal act or where the premises are insecure for an unknown reason. Where the owner or occupier of premises are unavailable then this creates a problem for the attending agencies, particularly the fire and rescue service and police service. Many FRSs and police services have arrangements with contractors to secure premises for which payment is then recovered from the owner/occupier or their insurers. Documentation should be left at the property explaining what's happened and what to do next including leaving a telephone number for any absent owners to contact the police/FRS for further information.

Where this is not possible, then every effort should be made to make the premises as secure as reasonably practicable and then handover responsibilities for the premises' security to the police service.

Where the owner/occupier is available then a complete handover should be given to them which includes details of the incident, hazards on the site, contact details of all relevant FRS staff and a written acknowledgement of the briefing and handover of the premises should be signed and a copy retained by the occupier. At this point, ongoing responsibility for the premises is accepted by the signatory and the FRS can close the incident.

Investigations

At many incidents, the extinction of the fire will not be the end of the FRS's involvement and the incident commander is responsible for initiating a number of actions and investigations.

Fire investigation

While some services automatically mobilise a fire investigation officer (FIO) to incidents that hit certain trigger points, it is nonetheless important that an IC confirms that a FIO is mobilised as soon as a crime is suspected or a fatality has occurred. An early understanding of fire dynamics can assist an FIO in determining the cause and spread of a fire, or identifying suspicious circumstances surrounding the incident. Close liaison between the IC and FIO will both help the investigation and reduce the impact of the incident on the FRS by co-ordinating resources to support the investigation, collating witness statements and ensuring full information is provided as quickly as possible.

Fire preservation of scene for criminal acts

At incidents where arson or other crimes are suspected, preservation of the scene is vital. The IC can support the police service by briefing crews, sector commanders etc of the need to preserve the scene and protect any evidence that may be essential. An early suspicion about the cause of a fire should prompt the IC to inform all personnel of the need to take care when working in the area to avoid unnecessarily moving materials and to be aware that they may potentially be operating in a crime scene. Where deaths are involved, advice should be sought from the police whether a body should be removed from a scene of operations or left in situ. When casualties have been located, where there may be even a slight suspicion that life may not be extinct, every effort should be made to carry out a rescue. If a casualty has been located but appears obviously dead, then the advice of the IC should be sought as to whether the body should be recovered or left in situ to assist the investigation of the incident. This may require the crew members to make a record of the details of the casualty – location, attitude, unusual aspects of the room layout etc – and may involve taking photographs using a thermal imaging camera or other recording device.

With commercial or even domestic premises, it may be the case that cameras or other image data capture devices are in use for security purposes. Where it is suspected that a crime has been committed, the police should be notified immediately and every effort should be made to protect this potential data from destruction by fire, and should be removed by others (preferably following the a advice or instructions of the police). Where such equipment has been removed in an emergency, it should be sent to the appropriate agency as soon as possible and a receipt received. Again, the IC should assist the police by making firefighters and other staff available for interview and/or prepare and collect witness statements to support investigations.

Fire safety: investigation of breaches of legislation and business safety issues

Where there has been breaches of the RR(FS)O, a fire safety officer will be required to collect evidence and interview owners/occupiers and other witnesses to determine whether to prosecute. It is also possible that lessons may be learned from the fire that could be applicable to other premises of a similar type or premises within a specific location. These business safety issues could be related to arson or accidental causes and are important for the purposes of ensuring business continuity and providing a warning to other premises.

Initiating community safety activity

Incidents, particularly in residential properties, are a good opportunity to promote domestic fire safety within the local area. ICs should be aware of this opportunity and initiate community safety/safe and well visits through the relevant department in the FRS. A 'hot strike' maybe possible within the immediate vicinity of a domestic incident, sometimes within hours of the fire occurring while circumstances are at the forefront of residents' minds. Co-ordination of such 'campaigns' is vital and an early indication to the relevant community safety department and/or other organisations, such as social services, health and education departments, will be of use. It may also be the case that as a result of the incident the IC becomes aware of vulnerable children and adults for which there may be safeguarding issues, and which will need to be raised through the relevant channels identified in service guidance and policy.

Again, many of the issues identified above are significant as associated with potential legal pitfalls. It is important therefore that these decisions are logged and rationales recorded for potential future use in legal hearings.

Post-incident staff welfare

It is important following difficult incidents where firefighters may have been exposed to extremely arduous circumstances, casualties or fatalities, that appropriate arrangements for post-incidence welfare checks are put in place. Normally, FRSs have procedures already in place and these should be used. It is possible that personnel issues may only become apparent after the incident, such as during debriefs or post-incident discussions and the IC must be aware of this potential occurrence.

Procedures, policies and equipment issues

Where it has become apparent that there are defects in policies, procedures and equipment, these should be raised through normal service routes or through debrief procedures (see Chapter 14). If there has been a major issue, then the IC should make sure that any actions required are followed through and not left to stagnate, particularly a where the outcome could have a significant impact on personal safety.

Cost recovery

The costs of some incidents may be recoverable from the owner, occupier or another person responsible for the incident. All services have cost recovery procedures which cover many circumstances, but other agencies for whom the fire and rescue service undertake work may be better placed to deal with the issues at some incident types. In particular, the environment agencies in the UK have very effective cost recovery teams and are adept at recovering costs using the 'polluter pays' principles associated with spillages and other hazmat incidents with environmental implications.

Summary

You should now understand the following concepts, techniques and practicalities discussed in this chapter:

The requirement to close down an incident in a controlled way ensuring that relief requirements are accurately estimated, and that transfers of command and reliefs are closely and safely managed.

The FRS should ensure that premises are handed over or secured responsibly to the owner or another responsible agency or organisation.

It is important to ensure that the requirement for carrying out post-incident investigations – fire investigation, fire safety inspections and scene preservation – are all completed and effectively managed.

The need to ensure that staff welfare is taken care of and any matters identified as a result of a debrief are forwarded for appropriate action.

Ensure that any potential for cost recovery is used and managed through the appropriate channels within the service.

Chapter 14: Debriefing and learning lessons

Introduction

It is a constant failing of the UK FRS to properly learn lessons from incidents of note – namely, those that have resulted in the deaths of firefighters or other significant losses due to inappropriate or incorrect actions undertaken by firefighters on the incident ground. There are a number of reasons for this including the relatively rapid turnover of more senior and experienced ICs who seem to leave the service without being able to transfer the legacy of their knowledge of incidents to their successors. There is also a relative unwillingness to share errors among other fire and rescue services, possibly for reasons of avoiding public or professional embarrassment. Some services have been brutally frank in publishing shortcomings from which other services could benefit; other FRSs avoid exposing their operational problems altogether.

One common failing on the incident ground is the failure to monitor the welfare and health of staff, particularly during times of arduous operational activity – it is a regular occurrence that firefighters have died or suffered injury as a result of heat exhaustion or exposure to heat and fumes. Despite these matters being referred to in the subsequent coroners' inquests and, subject to recommendations, these events still occur with a worrying frequency. In the UK, the National Fire Chiefs Council (NFCC) has recently introduced a National Operational Learning scheme whereby fire and rescue services share operational learning which has been in place since 2017. At the time of writing, the number of action notes based on issues of significant learning was four in some 18 months! This underlines the fact that organisational learning in the UK remains 'under the carpet', but it is hoped that critical incidents and vital lessons that need to be learned by the whole of the FRS (not only in the UK but in other countries) will be shared more freely in future.

Nonetheless, individuals, teams and even fire and rescue services have the ability to undertake internal learning. Interestingly, if a service identifies a critical failing but fails to promote it due to the reasons stated above, a second event of a similar type within that organisation (or presumably an organisation that has been made aware of the critical failing but had taken no action) will attract even more attention from health and safety enforcing authorities. Furthermore, it is likely to attract even greater fines or punishments than the initial transgression. For moral, legal and financial reasons, organisational learning should be embedded in every fire and rescue service and a systematic approach to learning lessons should be developed, and the findings, more importantly, should be disseminated among services across the country, rather than being 'kept in-house'.

Debriefing

Debriefing is an essential part of closing the learning loop from every incident. While incidents may have similar characteristics, even the simplest incident may have lessons that could assist firefighters and ICs in dealing with the next one. Where incidents of greater magnitude occur, then it is likely that many lessons could be learnt, including both aspects good practice and gaps in capabilities.

There are a number of different debriefing tools that are available to fire and rescue services to allow them to capture lessons from incidents, and the method of debrief selected will depend on the nature of the incident: its size, complexity, the nature of casualties, or a host of other aspects that may be of interest both within and without the service. Debriefs may be called different terms in different organisations, and there are several different types that may be used:

- Hot debriefs.
- Structured or formal debriefing.
- Incident command reviews.
- Incident monitoring reviews.
- Review of use of operational discretion.
- Multi-agency review.
- Critical incident stress debriefing.
- Personal reflection and development.

Hot debriefs

A hot debrief is one that is carried out as soon as possible following the conclusion of an operational incident (or exercise). A hot debrief is typically carried out when a serious incident occurs involving between two and five appliances. Ideally it should be carried out at the location of that incident and include all personnel who were involved. Representatives or participants from other organisations may also be asked to take part in the debrief, although this may stifle the debate that may be necessary among FRS staff. Where other agencies are used, they can bring a different perspective to an incident which may be of beneficial learning to all members of staff involved. The debrief should be conducted by the IC where possible and notes made of all relevant points that may be necessary to inform future operational decisions and procedures. It is important that the IC does not take the notes as this will distract him/her from engaging with firefighters and listening to what they have to say.

The structure of a hot debrief should include an overview of the incident from the IC's perspective to help crews understand what decisions were taken and why they were taken. It should consider the roles of the individuals and take account of their perspectives of the incident too, including things that went well and things that went less well. The tactical measures selected, risk control measures implemented as well as the operational equipment used should be reviewed to ensure that any new techniques or limitations of equipment are noted and passed on to the relevant department. Particular note should be paid to high-risk activities and how they were implemented, as well as the hazards that were present at the incident. Learning points should be gathered from all participants where possible and shared with them on completion of the post-incident report.

The advantages of a hot debrief are:

- It can be completed anywhere on the incident ground as long as it is outside the hazard area.

- It is immediate and can capture the contemporaneous thoughts of all participants and can have an immediate impact on the operational capabilities of staff.

- It is relatively quick to complete and requires a minimum of organisation at an operational incident.

- A hot debrief can include other agencies and helps cross-organisational learning for those attending.

- Equipment that has been used and records that are available (e.g. BA control boards, incident command packs etc.) can all be brought forward to demonstrate points of relevance.

- Any occurrence of recognition primed decision-making (RPD) (see page 102) may be remembered by those responsible for it and useful lessons on its application captured.

- As the debrief is taking place at the actual location, clarification of points and aspects of the incident may be cleared up immediately by walking to those relevant areas.

■ It is valuable in promoting good discussion and developing a wide understanding of decisions made within the context of the incident along with an explanation of the rationale, which may aid the understanding of everybody present and reduce the potential for adverse comments regarding the management of the incident.

Some of the disadvantages of a hot debrief are:

■ Preparation time for carrying out a hot debrief is limited as it is likely that crews will need to leave the incident scene to service equipment, refuel appliances and, in some areas highly dependent upon on-call firefighters, to return to the station so they may return to their full-time employment. This means that an officer in charge may not have sufficient time to prepare a debrief or have the time to carry one out. The consequence is that a hot debrief may be of limited duration and content.

■ It is possible due to the size and incident, that the IC, crew or watch commander may have limited training or experience of carrying out an operational debrief. As a result, lessons may be missed and not recorded.

■ At small incidents there is a tendency for the debrief be focused on task-level activities, and the wider operational and organisational learning may be missed, particularly if the debrief is carried out by someone who is not familiar with the requirements of higher levels of command.

■ The timing of the debrief is important – carrying one out before equipment is made up reduces availability of appliances. Readying appliances for redeployment may delay the debrief for a time and it may be the case that appliances are mobilised to other incidents in the meantime.

■ Carrying out an effective brief takes skill. Gaining the trust of contributors and encouraging those who are normally quite quiet and reserved may be a challenge and it is important that all participants feel they can talk about actions, omissions etc. in the presence of their peers, without fear of humiliation or embarrassment.

■ There is limited time for individuals to carry out personal reflection, and as a result some points being raised as learning issues may be perceived as bullying, victimisation or personal humiliation.

■ The lack of a trained facilitator could lead to negative feelings, interpersonal conflicts or the disengagement of participants.

■ It is important that any issues that have been identified are followed up because failure to do so may lead to the process being discredited and generate a lack of trust between managers and ICs.

■ Any comments may potentially be heard by none fire and rescue service staff in the vicinity and, crucially, it may be that not all those present have the best interests of the service in mind.

As can be seen, even a relatively simple process such as hot debrief requires skill and patience, as well as an understanding of the process by both crews and ICs.

Structured or formal debriefing

A structured debrief normally takes place when:

■ an incident has required a specified number of appliances (usually six as a minimum)

■ there have been a number of fatalities

■ complex rescues were needed

■ decontamination was required

■ it was an incident of special interest

■ it took place in a high-rise building

■ sleeping risks are involved, such as at hotels, hostels and guest houses.

The debriefs are planned to take place at a later date and away from the incident ground to allow a period of reflection and investigation before the debrief takes place. The organisation of a structured debrief can be complex and requires the use of a nominated organiser, who is responsible for making all the necessary arrangements, collating information, and organising the attendance of relevant personnel.

Due to the complexity of an incident and the potential consequences of any findings, it may be necessary for additional personnel to attend the debrief to add a specialist perspective and to provide technical advice to the facilitators and participants. There are a wide range of specialists who could be invited, including representatives of other agencies, fire safety officers and fire investigators, training personnel, representatives of blue light responders and fire control operators.

During the debrief, a debrief facilitator who may be specially trained will lead and facilitate discussions and manage the process. In this they will be aided by a sequence of relevant people who will develop the incident timeline and raise key issues for discussion. The debrief facilitator will ensure the collation of information and submission to the relevant responsible officers for consideration and, if appropriate, the implementation of any recommendations.

Issues covered during a structured debrief might include:

■ the timeline of the incident and key actions/omissions

■ the appropriateness of the command structure and the effectiveness of incident command

■ operational practices and procedures

■ the use of equipment

■ the communications structure, usage and effectiveness

■ the effectiveness of inter-service liaison and adherence to JESIP principles

■ external relations with the media, the public, business representatives and other partners involved in managing the incident.

Here are some things to bear in mind when conducting a structured debriefing:

■ Take a structured approach to breaking down the incident into stages of time.

■ A good facilitator will ensure a constructive and controlled debriefing environment.

■ Ensure there is ample time to prepare for the debriefing, including assessing and examining key issues that need to be raised and investigated in detail. Part of the preparation will include structuring the debrief to ensure that all key staff and agencies can attend and make contributions. Those that are unable to attend may be able to complete an incident debrief form to raise those points that they considered pertinent.

■ A neutral venue should be chosen as this will reduce the potential for particpant apprehension of a large structured debrief in a headquarters-type building. It will also help reduce the potential for disruption.

■ The interval between the incident and the debrief should be long enough to allow individuals to discuss the incident and reflect on their part and on the parts of others in resolving that event.

■ Allow those staff to have an influence on improvements to equipment, procedures etc.

Some of the advantages of a structured debrief are:

■ One of the key outcomes of the debrief is a written report with conclusions and recommendations for implementation to improve operational performance.

■ An effective debrief may create the opportunity for staff to identify potential development needs, which can feed into a personal development plan.

- An open and honest debrief can enhance the reputation of the service (and managers) as one that encourages open dialogue and seeks positive change and constant improvement.

- They provide an opportunity to discuss key issues, constructively challenge in a non-threatening environment, and where appropriate they celebrate success and good practice.

A structured debriefing is a large undertaking for an organisation and they can be time-consuming, expensive (particularly if on-call staff are required to attend), they can reduce the availability of resources and generally have an impact on productivity.

The time between the incident and the debrief, while good for allowing reflection, may mean that some details are forgotten, and this is why the immediate capture of data and feedback from a hot debrief or through electronic data capture is important and should be completed as soon as possible following an incident.

It is important that the facilitator is able to get the most out of staff. While the debrief is there to scrutinise what has taken place, it should not feel like an investigation or a 'trial' of the events. For this reason, it is important that senior and principal managers are seen to visibly support the debrief process, emphasising that the purpose of the debrief is to learn lessons and not to blame staff for any errors that might have occurred.

Incident command reviews

This type of review is a specialist form of structured debriefing that considers the operational aspects of an incident with a view to learning lessons and improving future command performance. It is important that, while the review should be undertaken by an officer more senior than the IC, it should take place in a no-blame and constructive manner, and aim to find improvements. Key decisions and actions that took place at an incident should be evaluated to ascertain their effectiveness and appropriateness for the circumstances. It may be necessary to enquire about the circumstances under which operational discretion has been used to find out whether such measures were necessary and valid, in which case it may be necessary to identify any policy gaps that need to be filled.

It is important that command reviews do not take place within an environment of secrecy as this may lead to unhelpful speculation by those not involved. Where possible, any lessons learned and proposed changes to operational procedures should be communicated as widely as possible.

Incident monitoring reviews

Many fire and rescue services now undertake performance reviews at operational incidents, sometimes focusing on general aspects of incident management and at other times focusing on specific issues, such as breathing apparatus operations, incident ground safety or command procedures. The outcomes of these reviews can be very valuable as they can produce comprehensive analyses of incident issues independently and contemporaneously. If properly conducted, they will be seen by all parties as valuable tools for learning lessons and promoting continuous improvement. The downside can be that they are viewed as 'spying operations' and can engender apprehension and concern by those being observed.

Reviews of operational discretion

Operational discretion can be a contentious issue and it is right and proper that any occurrence of its use should be properly reviewed to ascertain if it has been used appropriately. In these cases, evaluation of a service's procedures should be undertaken to identify how the gap in operational procedures can be effectively managed in future. If operational discretion has been applied incorrectly, then there may be training needs that need to be addressed by an individual, a team or the whole service. This review should take place in a less formal environment than a structured debrief but essentially covers the same issues in terms of the sequence of events, key decisions taken, outcomes achieved, and the decision-making process that led to the use of operational discretion.

Multi-agency review

When an incident involves large numbers of personnel from other organisations, a multi-agency review may be more appropriate than a single service review. The complexities of gathering all personnel who took part in the event together in the same room at the same time are generally too difficult to overcome and it may be necessary to undertake smaller reviews within each agency and collate the key findings to take to the multiagency review. Because of the difficulties and potential problems, there may be diverging opinions about what occurred and the appropriateness of the actions undertaken. It may be necessary for the debrief organiser and facilitator to be from an organisation that wasn't involved in the event. Potentially this could be somebody from the private sector with no obvious bias or links to the organisations involved.

Critical incident stress debriefing

Because of the nature of firefighting, there may be times when personnel are exposed to traumatic events. Critical incident stress debriefing (CISD) is a mental health management process with the goal of preventing adverse health outcomes and enhancing the well-being of employees. It is important that organisations seek the means to prevent the development of chronic and disabling problems such as post-traumatic stress disorder (PSTD), depression, substance misuse and relationship difficulties. CISD is a structured process whereby groups review a critical incident and their experience within that incident. It is designed to take place in a small group format and may be incorporated into a larger crisis intervention system. It is important that CISD is only carried out by experienced, well-trained practitioners and that it is non-mandatory. It should aim to support staff by helping to screen those at most risk, disseminate education and referral information and to improve organisational morale. This is a specialist type of debriefing and should not be undertaken by those without the necessary skills and training under any circumstances. **Note:** CISD may be known by othernames within organisations and while there may be differences in procedures, the principles generally follow a similar pattern.

Personal reflection and development

Finally, it is important that each individual taking part in an operational incident should undertake a period of reflection and consideration of the actions that they undertook, and that they critically evaluate those actions. It is expected of those firefighters who are on probation (or development) and those in development and promotion processes that they complete reflective logs of incidents they have attended and the activities they carried out. As part of a continual professional development process (CPD) and for reasons of professionalism, individual firefighters and commanders should undertake personal reflection to identify development needs on a regular basis. In that way, the degeneration from the competent firefighter/IC to the incompetent firefighter/IC due to a combination of skill-fade and overconfidence can be avoided, as a self-aware individual will identify any skills deficiencies and rectify them.

Other sources of information for learning lessons

There is a wide range of information that is available to all fire and rescue services to enable them to carry out an audit of operational skills and incident command effectiveness. Taken individually, each element may not give a complete picture of the situation but, in their totality, patterns may emerge that indicate knowledge, skills, procedural and information gaps which can, if not addressed, lead to difficulties in the future.

Some sources of additional information include:

- Corporate reviews.
- Workplace audits.
- Performance audits.
- Fire investigation reports.
- Periodic policy or procedure reviews.
- Health and safety reports.
- Accident reports.
- Near miss reports.
- Magazines and journals with relevant operational information that will assist in learning about new operational techniques, methods of building construction and lessons learned from other services.

Summary

You should now understand the following concepts, techniques and practicalities discussed in this chapter:

Appreciate the need for debriefs and their importance in improving personal, team and organisation performance.

Understand the advantages of the types of debriefs available to services: hot debriefs, structured or formal debriefing, incident command reviews, incident monitoring reviews, review of use of operational discretion, multi-agency review and critical incident stress debriefing.

Recognise the importance of personal reflection as a means to improve individual performance.

The availability of information from other organisations to support instant reviews and debriefs.

Chapter 15: Working together: Intra-operability and JESIP

Introduction

The legal framework that has formed the basis of the fire and rescue service in the UK for over 70 years has always included a provision for mutual aid and support to be given and received by all FRSs in times of crisis and emergency. These mutual aid provisions mean that a service requiring 50 or 100 pumps will be able to receive them. Other resources are becoming increasingly shared across service boundaries, and include officer cover and specialist resources such as urban search and rescue teams, fire investigation specialists and a range of others. Intra-operability at a fire service organisation level has been tried, practised and implemented almost on a daily basis, but using the resources of other fire and rescue services is not without risk and ICs must always be cognisant of the differences between services as well as the similarities.

Recently (since the attacks on the World Trade Centre in 2001 and subsequent acts of terrorism in the UK and elsewhere), interoperability between the various emergency services has become more important, partially in recognition of the fact that many incidents can only be resolved by close collaborative working and by sharing resources, ideas and co-ordinating actions to deliver a more effective service to the community. The Joint Emergency Services Interoperability Programme (JESIP) was introduced in 2012 to improve the joint working approach to incident management and has improved this collaborative approach to problem solving.

This chapter looks at the implications of working with other organisations and identifies how the IC can use the protocols for inter-service and inter-agency working to best effect.

Intra-operability: mutual aid

Interoperability is defined under the Fire and Rescue Services Act (2004) – sections 13 to 15. Under this act, a fire and rescue authority must enter into reinforcement schemes with other such authorities (s.13). They may also have arrangements with other organisations that employ firefighters, such as industrial organisations with equipment and resources for firefighting and airport fire and rescue services, inacluding defence premises firefighting staff and equipment (s. 15). Where arrangements are in place for other services to provide fire and emergency cover in another authority's areas – for example, a town near the border between two services may have its cover provided by the 'over the border' service rather than that which notionally provides cover within its own administrative boundary.

While the ability to call in large numbers of 'over the border' (OTB) resources is undoubtedly a great advantage, an IC should be aware of potential pitfalls that may occur and pre-empt those by ensuring that every effort is made to ensure that support and resources from other services are integrated into their own overall command structure. Some issues that may cause problems are set out below. Most of these are covered by specific arrangements, such as mobilising agreements and resourcing support, but familiarity with processes from both organisations may not be as full and complete as would be ideal.

Variable mobilising

The crewing of FRS vehicles is something that each service will determine and dictate for itself. Some services now crew a pump with only four firefighters, and the increasing use of smaller firefighting vehicles, sometimes called rapid response vehicles, means that crews of only two or three may attend incidents. In order to ensure that the number of resources requested is met correctly, the IC, when formulating assistance messages, will need to make clear to fire control the number of pumping resources, special appliances and the number of firefighters required, including the number of BA wearers. This will enable fire control, when mobilising resources and requesting mobilising of OTB resources through a different fire control, to specify what is required in such a way that the exact requirements of the IC are met.

Lack of information being received by OTB resources

Where OTB resources are mobilised, it may be the case that the information they receive may not be as complete as it should be. While the important thing is for them to attend the incident as quickly as possible, additional information may not be passed to them. It is necessary, therefore, for the IC to ensure that the briefing they receive upon attending is complete and possibly more intensive than that given to 'home crews', as they may have received additional information en route to the incident.

Equipment

Appliance equipment, specifications and capabilities vary, and before tasking OTB crews, it is important that a common understanding is achieved about any variations in equipment which may have an impact on the ability of those crews to complete the task. The IC should, wherever possible, provide a liaison officer to work with OTB crews to ensure full integration is being achieved and that any issues that need clarification can be addressed as quickly as possible. Fire ground radio communications may not always be compatible and command radios should be given to OTB sector commanders, BAECPS and BAECOs where necessary to ensure that communications are fully integrated.

Liaison and welfare

Where substantial resources from other brigades are dispatched to an incident, it is normally the case that a liaison and welfare officer is sent to support the crews and help integrate operations. This facility is not always part of an organisation's mobilising procedures and where additional officer support is required, a specific request should be made via fire control. A liaison officer from another service should be party to all relevant discussions that take place, so that OTB firefighters will be kept informed of progress. The liaison officer will also be able to help organise relief crews and manage the welfare of their crews.

Policies and procedures

Different services may have different policies for operational matters. For example, some authorities will not allow a firefighter who has already worn BA during that day to undertake a second BA wear. This can have serious implications at an OTB incident where additional resources are requested for breathing apparatus operations and the reinforcing appliances have insufficient BA wearers available. Planning should have taken this into account, and those stations in frequent receipt of cross-border resources should be aware of such limitations. It is important that the request for assistance is accompanied by a clear reason for those resources, especially when BA is required.

National Coordination and Advisory Framework (NCAF – England)

Most incidents to which FRSs respond are dealt with using local FRS resources. However, there may be times when a service requires specialist equipment or extra assistance to mount an effective response. An example would be a declared major incident or where extended deployment of resources may be necessary, or the deployment of specialist national resilience resources are required on a large scale. The National Coordination and Advisory Framework (NCAF) is part of the mechanism to provide the co-ordination of fire and rescue assets.

NCAF functions

NCAF has a wide remit for supporting local FRSs and communities through a number of co-ordination and deployment functions. These include:

- Mobilisation, co-ordination and monitoring of national resilience assets. Specialist assets and skills that are an integral part of the National Resilience (New Dimension) Programme.

- Chemical biological radiological nuclear (explosive) (CBRNe including detection, identification and monitoring (DIM) and mass decontamination).

- Urban search and rescue (USAR).

- High volume pumps (HVP).

- Enhanced logistics support (ELS).

Assets and skills hosted at a local level by fire and rescue services, or other agencies, that can respond if specifically requested to do so as a national capability and where incident timescales allow include:

- Water rescue.

- Marauding terrorist attack (MTA).

- Conventional and other specialist appliances.

- Bulk foam.

These assets and capabilities are located across a large number of fire and rescue authorities.

Strategic Holding Areas (SHAs) for managing the logistics and strategic deployment of resources to major incidents.

Home Office operations centres, which can be established in multiple locations and consist of government department officials and liaison teams, will provide situational awareness for use by COBR and other government bodies. They co-ordinate advice for ministers and engage with government liaison officers via the Resilience and Emergencies Division (RED), which is part of the Ministry For Housing Communities And Local Government (MHCLG). They provide strategic co-ordinating groups with a single point of contact for central government assistance.

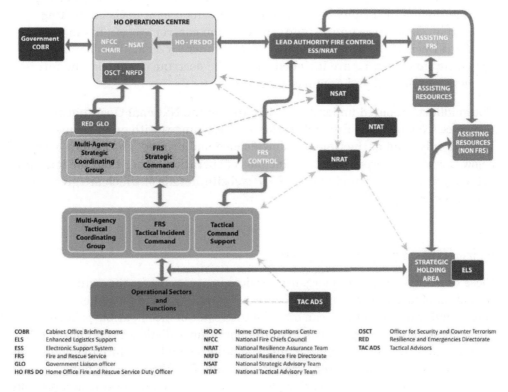

COBR	Cabinet Office Briefing Rooms	HO OC	Home Office Operations Centre	OSCT	Officer for Security and Counter Terrorism
ELS	Enhanced Logistics Support	NFCC	National Fire Chiefs Council	RED	Resilience and Emergencies Directorate
ESS	Electronic Support System	NRAT	National Resilience Assurance Team	TAC ADS	Tactical Advisors
FRS	Fire and Rescue Service	NRFD	National Resilience Fire Directorate		
GLO	Government Liaison officer	NSAT	National Strategic Advisory Team		
HO FRS DO	Home Office Fire and Rescue Service Duty Officer	NTAT	National Tactical Advisory Team		

Figure 15.1: The NCAF in operation
(Reproduced from NOG)

Interoperability: multi-agency working and JESIP

Integrated working by the emergency services at major incidents has been common practice for many years. Co-ordination of the response has generally been effective and helps communities cope with disaster and recovery. Following a series of major incidents where interoperability was less than optimal the Derrick Bird shootings in Cumbria 2010, the London bomb attacks in 2005 and wide area flooding in 2005 and 2007) three emergency services – police, ambulance and fire and rescue – developed a programme to improve inter-agency working. The Joint Emergency Services Interoperability Program (JESIP) was developed in 2012 with the intention of ensuring that all *'blue light services are trained and exercise to work together as effectively as possible at all levels of command response to major or complex incidents (including fast-moving terrorist scenarios), so that as many lives as possible can be saved.'*

The programme sets out the approach to dealing with major incidents and details principles for joint working, arrangements for joint working, sharing information and principles for action incorporating a joint decision-making model that all services understand. The decision control process fits in with the Joint Decision Making Model and is complimentary. Facilities for cross-organisational learning are included within the doctrine to support the future development of staff and organisations.

The guidance prepared under JESIP, like that of the National Operational Guidance (NOG) for the fire and rescue service, changes with great frequency and it is only valid at the time it is downloaded and printed. The guidance set out below therefore incorporates the principles of operation rather than the technical details, which can be found at the JESIP website, details of which are given in the references section of this book.

Principles for joint working

Close working with other emergency services is necessary for the swift resolution of many operational incidents. Sometimes working together has been found wanting but this can often be the result of incompatible policies, misunderstandings of operational requirements, unrealistic expectations of other agencies' abilities and capabilities, and sometimes simply a failure to understand the command structure for the types of incidents. The JESIP programme identified a number of simple principles that underpin joint working and promote good practice. These principles should be introduced as soon as possible at the early stages of an incident where more than one agency is involved, and it should be retained throughout the duration of the incident.

The five principles are:

- **Co-locate:** Where ICs are all within close proximity to each other, direct and face-to-face communication is possible. As is well known, words are far more effective when one can see the expressions on people's faces and the gestures they make, and this direct contact will improve the transfer of information and knowledge between agencies. As soon as possible during an incident, the senior commanders from all agencies should seek the others out and agree a location for the forward command post (FCP). Each service commander should wear the IC tabard for their service.

- **Communicate:** All communications should be clear and jargon free where possible, using plain English (see Chapter 12 – communications).

- **Co-ordinate:** Agree the lead service and identify priorities, resources and capabilities for an effective response. Ensure that the expectations of other services are realistic. Organise a meeting schedule that all can manage.

- **Joint understanding of risk:** complete a joint risk assessment to ensure understanding of the potential impact and possible control measures.

- **Shared situational awareness:** develop an understanding of the situation through use of the Joint Decision Making Model, complete a METHANE message (see page 255), and the preparation of an IIMARCH briefing document (see page 257).

Incident commanders from all agencies involved should try to arrange a joint meeting as soon as possible, ideally within 30 minutes of arrival at the incident.

Joint strategy development

Where these principles are followed, a strategy should be prepared that sets out specific actions for each agency. The strategy should address the following issues:

- **What** are the aims and objectives to be achieved?

- **Who by?** Police, fire, ambulance and partner organisations?

- **When?** Timescales, deadlines and milestones.

- **Where?** What locations?

- **Why?** What is the rationale? Is this consistent with the overall strategic aims and objectives?

- **How?** By what means are these tasks going to be achieved?

The decision support tool for assisting with multi-agency incidents is the Joint Decision Making Model (JDM) and should be used to deliver a strategy with clear aims and objectives.

The Joint Decision Making Model (JDM)

Use of the JDM will help collate all available information, clarify joint aims and objectives and help make robust and effective decisions. There are three primary considerations:

- **Situation:** What is happening? What are the impacts? What are the risks? What might happen and what is being done about it? This relates to the acquisition of joint situational awareness and the generation of possible solutions.

- **Direction:** What outcomes should be achieved within the first hour? What are the aims and objectives of the emergency response? What overarching values and priorities will inform and guide this?

- **Action:** What needs to be done to achieve the outcomes?

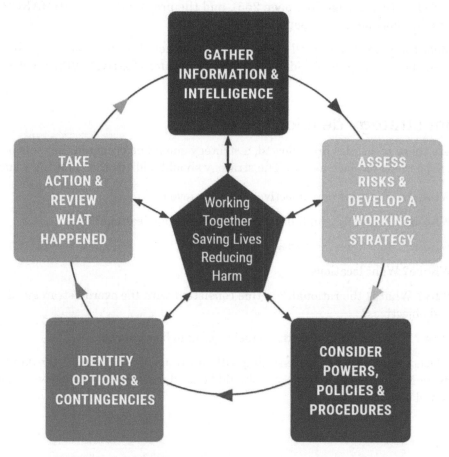

Figure 14.2: The Joint Decision Making Model
(Reproduced with permission of JESIP)

The JDM should be used in a multi-agency context with all parties making a contribution to setting the outcomes, aims and objectives. Like most decision support tools it is iterative, and intended to be reviewed regularly and as circumstances change. The DCP supports the JDM and feeds information into the JDM by 'Assessing risk and developing working strategy as shown below, supporting the FRS aims as well as the overarching multi agency strategy'. There are a number of key elements to the JDM that enable the agencies to work together to reduce harm and save lives. These elements are briefly detailed below.

Gathering information and intelligence

The first stage is to gain situational awareness through discovering what is happening, what the impacts are, what the risks are, what might happen and what is being done about it. The sharing of information with partners will help produce a robust picture of what is going on and what could happen. It is also important to identify what is not happening and to gain a common understanding of the aims of other organisations and their capabilities and limitations. In order to assist the development of a shared understanding of the situation, the JESIP information sharing model, known as METHANE, may be used.

The METHANE model has been used for a number of years as a reporting framework providing a common structure for responders and their control rooms to share major incident information. (It may also be used for other incidents falling below the major incident threshold with the omission of the 'M', to become an 'ETHANE' message.) Determining if an incident merits the declaration as a major incident should be completed as early as possible, the declaration made and a METHANE message sent to all service controls as soon as possible. The first agency to arrive on scene should send the M/ETHANE message so that situational awareness can be established quickly.

M   Major Incident declared?

E   Exact Location

T   Type of incident

H   Hazards present or suspected

A   Access – routes that are safe to use

N   Number, type, severity of casualties

E   Emergency services present and those required

Figure14.3: The METHANE model for sharing information
(Reproduced with kind permission of JESIP)

Assess risks and develop a working strategy

Assessing risks and developing a working strategy ensures that commanders have reviewed and assessed risks so that control measures can be put in place. Understanding risk is central to an integrated emergency response and responders must continually expand and maintain a shared understanding of the full range of risks, and the impact of their actions on those risks. The DCP dovetails into this element and supports strategy development.

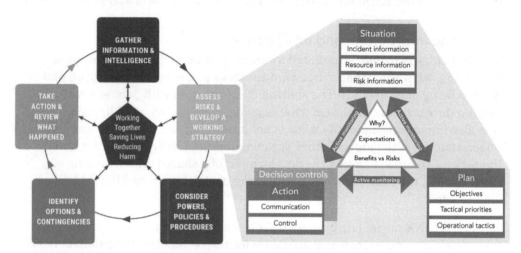

Figure 14.4: JDM and DCP interrelationship
(Reproduced with kind permission of JESIP)

Consider powers, policies and procedures

Proposed actions must always comply with the legal powers and duties of the responding agencies and this step is to ensure that no laws, including Human Rights laws, are threatened by the response of the emergency services. Considerations should include:

- What relevant laws, standard operating procedures and policies apply?

- How do these influence joint decisions?

- How do they constrain joint decisions?

A common understanding of any relevant powers, policies and procedures is essential in order to avoid conflict between the actions of one service and the legal duties of another.

Identify options and contingencies

Once options have been generated, they should each be rigorously assessed for their suitability, feasibility and acceptability. Once determined, clear plans of action should be drawn up and include communications procedures for deferring or aborting the actions and also to initiate a specific tactic.

Take action and review what happened

The JDM is an iterative process and the review stage forms part of the process. The outcomes of actions, new information and changes in situational awareness all feed into the review stage, which then allows actions to be refined, revised or changed to meet the needs of the evolved incident.

Sharing situational awareness: IIMARCH

As incidents develop, the briefing tool IIMARCH should be used by civilian agencies, with information briefed against each heading in the IIMARCH mnemonic (**I**nformation, **I**ntent, **M**ethod, **A**dministration, **R**isk assessment, **C**ommunications, **H**umanitarian issues). However, in the early stages, a briefing can be delivered quickly around the content of the JDM or DCP.

I	Information
I	Intent
M	Method
A	Administration
R	Risk Assessment
C	Communications
H	Humanitarian issues

Figure 14.5: The IIMARCH Briefing Model
(Reproduced with kind permission of JESIP)

Summary

You should now understand the following concepts, techniques and practicalities discussed in this chapter:

The concept of mutual aid, its legal basis and the difficulties associated when working with over the border resources.

The processes and functions of the national coordination and advisory framework within the UK and how it may assist by providing a national response to a local emergency.

The concepts of strategic holding areas (SHAs) and the operation of NCAF.

The need for interoperability in multiagency working within the UK concept of operations and the principles of joint working, strategic development and the use of the joint decision-making model (JDM) in delivering better outcomes for the community. Appreciate the interlinking between the JDM and DCP to deliver practical operational outcomes at incidents where the FRS are involved.

The METHANE and IIMARCH models for sharing information between agencies and government at major incidents.

Chapter 16: Major incident management – the UK Concept of Operations

Introduction

The management of major emergencies in the UK as a subject worthy of several books in itself and it is not the intention of this chapter to give comprehensive coverage of such a vast subject. It is, however, a subject of which all ICs should be aware. It is likely that in the event of a major incident occurring, a fire and rescue IC will be one of the first in attendance and knowledge of how the incident management organisation will develop will enable them to start to put a formative disaster management structure in place. This chapter will cover the principles of emergency management in the UK, government responses to major emergencies (including the government's supporting structures), the levels of emergency, the response at a local level and the responsibilities of emergency responders at these incidents.

Principles of emergency management in the UK: Concept of Operations

Emergency management in the UK has become more structured in the 21st-century and has been formed around the Civil Contingencies Act (2004). This act provides an inclusive definition of 'emergency', which is *'a situation or series of events that threatens or causes serious damage to human welfare, the environment or security in the United Kingdom'*. These events include a wide range of scenarios including adverse weather, severe flooding, animal diseases, terrorist incidents and the impact of a disruption on essential services and critical infrastructure.

It is important that this definition of emergency is not confused with the term 'major incident', which is defined as *An event or situation, with a range of serious consequences, which requires special arrangements to be implemented by one or more emergency responder agencies'*, with the key phrase being 'emergency responder agencies'.

The guiding principles for UK major disaster response are:

Preparedness: All organisations and staff that may have to respond to an emergency should be properly prepared, which includes having an understanding of their role and responsibilities, having prepared plans to respond to such emergencies, and undertaking rehearsals to ensure that the response is effective when required.

Continuity: The operations being undertaken as part of the response to an emergency should be based upon an organisation's response to 'normal' activities and ways of working. It is recognised that the scale of operations will be increased due to the nature of the emergency and will require a greater degree of control and co-ordination, particularly at tactical and strategic levels, but as far as the tasks that need to be carried out they will in essence be based around existing procedures. (See roles and responsibilities of emergency responders below).

Subsidiarity: All decisions should be taken at the lowest appropriate level and co-ordinated at the highest level necessary. In essence, this means that those at local level form the basis of a response to an emergency of any scale and duration.

Direction: Strategic aims and supporting objectives should be clear and provide the clarity of purpose for all those responding. These aims and objectives should be agreed by all involved in mounting a response at a strategic level and will aid those delivering outcomes by enabling them to prioritise and focus their response.

Integration: Integration and co-ordination of response should be exercised at all levels of government and organisation to ensure that an effective response is made to an emergency.

Communication: Reliable and effective communications that allow the transmission of mission-critical information in a timely manner is essential. The need to inform all levels of command as well as the public directly and through the media requires that communications are resilient under all conditions.

Co-operation: Organisations involved in an emergency response should positively engage with others and share situational awareness and information.

Anticipation: The identification of potential risks at all levels will allow planning to be undertaken so that the likely response can be anticipated and an understanding of how to manage the incident will be developed before the occurrence of an emergency.

The national response to serious categories of emergencies

Levels of emergency

Any emergency has a number of common phases:

- The preplanning or preparation phase.

- The response, which is dealing with the immediate issue or preventing escalation.

- Recovery, which involves the longer term activity of returning the community or area to its pre-emergency condition as far as possible.

The response phase is further divided into 'crisis management' and 'consequence management'. Crisis management involves preventing an imminent emergency and implementing interventions to mitigate the effects and to manage the focus of the incident. For the FRS, these emergencies may include events such as major fires or hazardous material spillages, wide scale flooding and wildfires, transport infrastructure incidents or aircraft crashes etc. Other incidents, such as terrorism-related events or major medical emergencies, may not seem to be the responsibility of the fire and rescue service, but resources may well be requested to provide support to other agencies.

There are five levels of emergency categorised by the government, although the government itself is not involved in the two lower levels – local (minor) and local (major) incidents. Central government does have a role in three top tiers of emergencies: level 1 (significant emergencies); level 2 (serious emergencies); and level 3 (catastrophic emergencies). These are defined as follows:

Significant emergency (Level 1): This is a serious incident which is normally dealt with at local level but requires support from central government. This is provided by the lead government department (LGD), or the relevant department within the UK devolved administrations, which is the department responsible for the area of activity in which the emergency has been declared. Thus, for a major incident involving a health issue such as an outbreak of measles in a small geographical area, the lead government department would be the Department of Health. For a terrorist-related emergency, the Home Office would be the lead. An accidental release of radioactive nuclear material in the UK would be responsibility of Department for Transport if in transit, or Business, Energy, Innovation and Science, and an overseas emergency will be the responsibility of the Foreign Office. These incidents will not require a collective central government response unless the event requires it to become necessary.

Serious emergency (Level 2): This emergency has or threatens a wide or prolonged impact that necessities support from central government, which will co-ordinate and provide support from a number of departments and agencies. This response would be co-ordinated from the Cabinet Office Briefing Room (COBR) and would be led by the lead government department.

Catastrophic emergency (Level 3): This is the most serious level of incident, which threatens an exceptionally high and potentially widespread impact. It would require immediate direction and support from central government. For this level of support to be required, the local response is likely to have become overwhelmed by the extent of the incident or the number of casualties. It may also be the case that emergency powers are required to be invoked.

The role of the Cabinet Office Briefing Rooms – COBR

The Cabinet Office is responsible for working with central, local and regional partners to prepare for emergencies and help co-ordinate central government's responses to major incidents. For level 2 and 3 emergencies, the Cabinet Office Briefing Rooms (COBR) provides a physical location from where the government's response is activated, monitored and co-ordinated. The relevant COBR committee will become the focal point of the national response and provide authoritative advice and guidance for local responders. The objectives of COBR will initially be protecting human life, property and the environment and to alleviate suffering; supporting the continuity of everyday activity in the restoration of disrupted services; and upholding the rule of law and the democratic process. Needless to say, once the specific details of an incident are identified, the strategic objectives will be refined and clarified.

COBR structure and supporting cells

A range of organisations and agencies will sit within the COBR structure and will be supported by working groups, or 'cells', which provide specialist advice and undertake investigations and activities to support that advice. These cells can include but are not limited to:

- **An Intelligence Cell,** staffed by the intelligence agencies, will co-ordinate information and surveillance data in the the event of an act of terror requiring COBR to stand up. Among other things, it will produce a high level intelligence assessment, preparing planning assumptions and co-ordinate COBR intelligence requirements.

■ **A Situation Cell**, which will co-ordinate information about the incident and develop a projection of likely developments of an event across a series of time frames. The Situation Cell develops and maintains a Common Recognised Information Picture (CRIP), which will be used to brief participants at COBR and those responders at local level (where possible – due to security restrictions). The CRIP will consist of information about the event and significant wider impacts, including facts and figures, the main developments and decisions, trends, and upcoming decision points. The Situation Cell will also provide the necessary information for briefing ministers.

■ **Operational Response** group will help lead and co-ordinate the response to an emergency situation and may be the responsibility of the LGD in the event of a non-terrorist event or by the Home Office in the event of an overt terrorist incident within the UK.

■ **Specialist Scientific Advice Cells** will be stood up to support decision making where there are public health or environmental implications from an incident. The establishment of a Scientific Advisory Group for Emergencies (SAGE) will occur when a decision has been made by the LDG of the Cabinet Office depending upon the severity of the actual or potential consequences.

■ **Specialist Legal Advice.** Major incidents invariably raise a number of serious legal questions and this specialist group will be established from the relevant department's legal teams to provide authoritative guidance to COBR.

■ **Logistic Support.** Wherever appropriate, logistical support will be managed at local level, but where additional support is required, this will be the responsibility of the LGD and a 'logistical Operations Cell' may be established.

In managing its liaison and support for the local response, COBR can make use of the Ministry for Housing, Communities and Local Government, Resilience and Emergencies Division (MHCLG – RED), which has the following functions:

■ Acts as a conduit for communications between central government and the local level.

■ They are responsible for supporting local responses and recovery efforts, and ensuring that there is an accurate picture of the situation in their area through close liaison with the strategic co-ordination group(s) (SCGs – see below) at local level.

■ If emergency powers under the Civil Contingencies Act are to be enacted, a nominated co-ordinator will be required. They will co-ordinate activities under emergency regulations.

The local response to major emergencies

For incidents that do not require the involvement of central government, that is, a more localised incident where there is nevertheless a requirement for the deployment of large numbers of resources from one or more of the emergency services and other category one and two responders, the structure of managing a large emergency in terms of tiers of response depends on the nature of the incident and the level of resourcing and support that is required to mitigate damage and disruption.

Strategic co-ordinating groups (SCGs)

The purpose of an SCG is to take overall responsibility for the multi-agency management of an emergency and establish the policy and strategic framework within which lower levels of command and co-ordinating groups will work. This includes:

- Determining and disseminating clear strategic aims and objectives, and reviewing them regularly.

- Establishing a policy framework for the overall management of the event or situation.

- Prioritising the requirements of the tactical tier and allocating personnel and resources accordingly.

- Formulating and implementing plans to handle the media and public communications, potentially delegating this to one responding agency.

- Direct planning and operations beyond the immediate response in order to facilitate the recovery process.

The SCG is formed by senior representatives with executive authority (i.e. they can spend money and deploy resources, equipment etc.) from each of the key organisations responding to an emergency. The SCG will normally work from a strategic co-ordination centre (SCC), which is usually in a police headquarters. During the initial response phase, the senior police officer will normally chair the meeting but this is not necessarily the case. The SCG will collectively determine the strategic decisions for managing the incident. It is important to remember that the SCG should not be referred to as 'Gold', as this may confuse matters because some organisations still use the term 'Gold' for internal command and management structures.

Parallel to the SCG, each agency will set up a parallel command structure to support the SCG and be led by the agencies' 'gold commander'. Fire and rescue services may set up an incident room from where their own strategic activities may be conducted – this function is often referred to as 'fire gold command'.

A typical SCG structure may include representatives from category 1 and 2 responders and be chaired by the most appropriate lead agency (for example, a Public Health England representative for a pandemic flu emergency, police for a terrorist related incident etc.). For a serious rail accident, the typical SCG membership will include Blue light services, PHE, the NHS, British Transport Police, Rail representative, Government Liaison Officer, the media and possibly military representation. The SCG is supported by a number of subgroups, which could include a recovery co-ordinating group (normally led by a local authority representative), a scientific and technical advisory cell (STAC) led by a relevant expert organisation (Public Health England or the Environmental Agency, for example), and it may be necessary to instigate the creation of a media cell, commonly led by the local police's press liaison team. Usually the recovery aspects of the incident work in parallel to the main SCG activities.

Strategic objectives

There are a number of common strategic objectives for a wide range of incidents and usually these can be predetermined and selected depending on the nature of the incident. These objectives include:

■ saving and protecting human life

■ relieving suffering

■ protecting property

■ providing the public with information

■ containing the emergency – limiting its escalation or spread

■ maintaining critical services

■ maintaining normal services at an appropriate level

■ protecting the health and safety of personnel

■ safeguarding the environment

■ Facilitating investigations and inquiries

■ promoting self-help and recovery

■ restoring normality as soon as possible

■ evaluating the response and identifying lessons to be learned.

Below the multi-agency SCG, the three levels of command for each responding agency will be implemented. These levels are strategic (gold), tactical (silver)

and operational (bronze) (see Figure 16.4). At many incidents the instigation of the three-tier command structure will take place without other agencies being involved and may be undertaken without reference to any other organisations. It is important, however, that if a major incident is declared by any category one or two responder, notification procedures are such that they will inform other agencies to make them aware that 'business as usual' may not be available for some time.

Figure 16.4: the 3 Tier Emergency Management Model

Other levels in the three-tier emergency management model

Due to the scale of incidents normally dealt with by firefighters, it is more likely that the lower levels of the IEM - the tactical and operational levels – are more familiar to them than the SCG level. Broadly speaking, the 'tactical level' equates approximately to an 'on site' command struture at larger or more significant and complex incidents where the incident commander is located in a command unit and is liaising with other service commanders at a tactical level. At incidents where a Tactical Co-ordinating Group (TCG) is established off site (usually at a local police divisional headquarters), this will operate in parallel to the tactical commanders group on the incident ground. An understanding of how this added complexity is managed in a local area or jurisdiction is necessary to enable effective command and co-ordination of these incidents. Pre-incident liaison and exercising with partners will aid effective implementation on the ground during the real thing.

The operational level of the model is the most easily recognised element as it occurs at all incidents with the incident commander (or operational commander[2]) having responsibility for the management and command of staff dealing with specific tasks which are used to resolve the incident. At larger incidents there

---

2   Not to be confused with the 'Operations Commander', who is tasked with co-ordinating and controlling several sectors.

may be several operational commanders and it is likely that they report to the on site 'Tactical Commander' who is the incident commander of a larger incident.

Responsibilities of emergency responders

During large emergency situations, it is likely that all three emergency services will be in attendance. The roles and responsibilities of responders have been clearly set out in government guidance (Emergency Response and Recovery, 2013).

Fire and rescue authorities

The primary role of a fire and rescue authority in an emergency is to

- extinguish any fires and rescue anyone trapped by fire, wreckage or debris
- prevent further escalation of an incident by controlling or extinguishing fires, rescuing people, and undertaking other protective measures
- deal with released chemicals or other contaminants in order to render the incident site safe, or to recommend exclusion zones
- assist other agencies in the removal of large quantities of flood water
- potentially assist ambulance services with casualty-handling, and the police with the recovery of bodies.

Note: in some areas there are agreements between fire and rescue and the police for controlling entry to cordons. Where this is the case, fire and rescue personnel are trained and equipped to manage gateways into the inner cordon and will liaise with the police to establish who should be granted access, and keep a record of people entering and exiting.

A fire and rescue authority may undertake mass decontamination of the general public in circumstances where large numbers of people have been exposed to chemical, biological, radiological or nuclear substances. This is done on behalf of the NHS, in consultation with ambulance services.

Police services

The police will normally co-ordinate the activities of those responding to a land-based sudden impact emergency, at and around the scene. There are exceptions however, for example a fire and rescue authority would co-ordinate the response at the scene for a major fire.

For the police, as for other responders, the saving and protection of life is the priority. However, they must also ensure the scene is preserved so as to safeguard evidence for subsequent enquiries and any criminal proceedings. Once life-saving is complete, the area will be preserved as a crime scene until it is confirmed otherwise (unless the emergency results from severe weather or other natural phenomena and no element of human culpability is involved).

The police oversee any criminal investigations. Where a criminal act is suspected, they must undertake the collection of evidence, with due labelling, sealing, storage and recording. They facilitate inquiries carried out by the responsible accident investigation bodies, such as the Health and Safety Executive (HSE) or the air, rail or marine accident investigation branches. If there is the possibility that an emergency has been caused by terrorist action, then that will be taken as the working assumption until demonstrated otherwise.

Where practical, the police, in consultation with other emergency services and specialists, will establish and maintain cordons at appropriate distances. Cordons are established to facilitate the work of the emergency services and other responding agencies.

Where terrorist action is suspected to be the cause of an emergency, the police will take additional measures to protect the scene (which will be treated as the scene of a crime) and will assume overall control of the incident. These measures may include establishing cordons to restrict access to, and require evacuation from, the scene, and carrying out searches for secondary devices.

All agencies with staff working within the inner cordon remain responsible for the health and safety of their staff. Each agency should ensure that personnel arriving at the scene have appropriate PPE and are adequately trained and briefed. Health and safety issues will be addressed collectively at multi-agency meetings on the basis of a risk assessment. If it is a terrorist incident, the police will ensure that health and safety issues are considered and this will be informed by an assessment of the specific risks associated with terrorist incidents.

The police process casualty information and are responsible for identifying and arranging the removal of fatalities. In this task they act on behalf of HM Coroner, who has the legal responsibility for investigating the cause and circumstances of any deaths involved.

Survivors or casualties may not always be located in or immediately around the scene of an incident. It is therefore important to consider the need to search surrounding areas. If this is necessary, the police will normally co-ordinate search activities on land. Where the task may be labour intensive and cover a wide area, assistance should be sought from the other emergency services, the Armed Forces or volunteers.

Ambulance services

As part of the NHS, Ambulance Trusts have the responsibility for responding to and co-ordinating the on-site NHS response to short notice or sudden impact emergencies.

They are responsible for identifying the receiving hospital(s) to which injured people should be taken, which, depending on the types and numbers of injured, may include numerous hospitals remote from the immediate area of the incident. The person with overall responsibility for this at the scene of an emergency is the ambulance IC (AIC).

Ambulance Trusts, medical teams and other emergency services endeavour to sustain life through effective prioritisation of emergency treatment at the scene. This enables the AIC to determine the priority for the release of trapped people, treatment, and where necessary the decontamination of casualties. This will allow patients to be transported in order of priority to receiving hospitals.

Ambulance services may seek support from other organisations, specifically the voluntary sector (e.g. British Red Cross, St John Ambulance), in managing and transporting casualties. Where deployed, these organisations would work under the direction of the Ambulance Trust.

Ambulance services also have Hazardous Area Response Teams (HARTs). These are highly trained staff with specialist equipment capable of entering hazardous environments with the sole purpose of saving life. Each Ambulance Trust has at least one HART and they are strategically located throughout the country.

Where there are large numbers of casualties, ambulance services may establish a Casualty Clearing Station (CCS) at or near the scene of the incident. The CCS is designed to provide medical care for those injured at the scene.

Summary

You should now understand the following concepts, techniques and practicalities discussed in this chapter:

The application of the UK concept of operations and its implications for the FRS and the principles of emergency management in the UK.

The national response to serious categories of emergencies which form the basis of UK resilience planning.

The role and structure of he Cabinet Office and in particular Cabinet Office Briefing Rooms (COBR) and its use during major emergencies.

The linkage between central and local response including the use of the strategic coordination group (SCGs), tactical coordinating group (TCG) and operational activities in the incident ground.

The responsibilities of emergency responders including the fire and rescue, police and ambulance services.

Chapter 17: Concluding thoughts: the fire and rescue service and the future

As we enter the post-Grenfell Tower fire world, in many respects the fire and rescue services within the UK are at a crossroads: reducing numbers of firefighters in totality and in crews on fire engines mean that the way incidents are managed will need to change; the evolving role of firefighters within their communities will add pressures on the service but increase community benefits; in governance structures including the merger of fire and police services under the police and crime commissioner's auspices may result in changes to the way services are managed and the way it works with other organisations; the way information is produced and accessed gives us all sorts of advantages but also leaves us at a disadvantage when changes occur too quickly; and eternal squeeze on public finances means that 'doing more with less' and 'working smarter not harder' (and several other platitudes) becomes a way of life rather than a one-off.

The number of firefighters riding frontline fire appliances in many services have now become standardised at four at all times. Some experimentation is being undertaken within other services of crewing rapid response vehicles (called various names by fire and rescue services) with two or three firefighters. This doesn't necessarily mean things are going to get worse but that incident commanders need to think in different ways. For example, the shorthand messages from the incident ground requesting additional resources will not necessarily be the most appropriate way of dealing with things in future – the 'make pumps 10' may become 'five pumps, 45 firefighters and six command officers' – or even today.

New roles for firefighters such as corresponding for medical emergencies should help secure a future for a service that has already shown that it is capable of reducing community risk at a national level. Supporting medical services in further reducing community risk through long-term health improvement programmes and, more immediately, in continuing emergency response to medical emergencies will undoubtedly need more developed medical skills in the future which will have an impact on the way firefighters are trained, organised and commanded. Additional demands such as these place a greater burden on the service and have the potential to reduce the global availability of fire appliances, and again will prompt thoughts about how the incident commander should manage these aspects in the future and prepare to do more with less, particularly in the early stages of incidents due to scarcer resources.

The educator role of the firefighter will change as civil emergency response for major disasters are likely to follow the trends being set in Asia where an 'active citizenship' is being encouraged (in places like Singapore) for communities to be trained to support themselves. The community responders will need training in disaster response and incident commanders will need to develop skills to lead civilian responders at incidents.

The changes in equipment and technology are also having an impact on the service. Ultra high pressure hoses and jets are expected to make a greater appearance as tools the firefighter uses at fires. Unlike the wholesale adoption of gas cooling techniques across the British fire service, following an incident at Blaina, Gwent in 1996, the service must avoid seeing new devices as a panacea for all fires. Rather they should be viewed as another tool in the toolbox, selected by competent firefighters and incident commanders, to meet the specific needs of the incident.

The new governance structures involving the police and crime commissioners (PCCs), implemented in many fire and rescue services are likely to bring closer collaboration and working practices between police and fire and this will likely as not improve all blue light service liaison. Similarly, the slow moving trend towards mergers between fire and rescue services and the sharing of operational resources (including command and specialist officers) across service borders is likely to continue, which creates a demand for integration both of policies and operational procedures.

The introduction of the national operational guidance programme is undoubtedly increasing the amount of information available to services and individuals with regard to operational procedures. At the time of writing, the continual churn and updating of information, sometimes on a daily basis, makes it problematic and overwhelming to ensure currency of understanding within organisations. One of the purposes behind this book was to provide a simple reference document with principles and good practice for incident command and to avoid getting bogged down in fads or temporary trends. It is our intention to periodically revise the contents of the book but not to amend it on an almost weekly basis.

The financial squeeze has been with the UKFRS for over a decade and has undoubtedly caused a huge challenge for the service. On a positive note, it has once again proved the adage that 'necessity is the mother of all invention' and many changes have allowed the service to take an evolutionary step forward in terms of technology, information and equipment. The practical downsides of slower attendances and fewer firefighters on first attendances are the trade offs for 'efficiencies'. The financial difficulties are not something that will go away and further evolution is inevitable.

What this means for incident commanders is that they must continually develop their own skills and knowledge of the service, the communities they serve and keep abreast of trends and changes within the profession. As ever, it means they must continually seek information about incidents, new equipment and new techniques so that they may apply their skills effectively where it matters: on the incident ground. Making time for self-development and reflecting on incidents that have been attended is something that many busy fire officers and incident commanders often put aside because of competing demands from other parts the role. They should remember that the safety of firefighters at incidents is paramount and that their leaders should be competent, knowledgeable and technically proficient if they are to command on the ground.

The future for any commander will be challenging, but time and effort spent learning and updating knowledge and skills will be repaid when everybody returns to their fire station fit and well, knowing they have effectively served their communities.

References

Selected bibliography

Adair J (2002) *Effective Strategic Leadership*. London: Pan.

American Psychological Association (2018) *Stress: The different kinds of stress, accessed* [online]. Available at http://www.apa.org/helpcenter/stress-kinds.aspx (accessed February 2019).

Blackston GV (1957) *A History of the British Fire Service*. Routledge and Kegan Paul.

Braidwood J (1866) *Fire Prevention and Fire Extinction* [online]. Available at: http://www.gutenberg.org/ebooks/26440 (accessed February 2019).

Brunacini AV (1985) *Fire Command* (1st Edn). Quincy, Mass: NFPA.

Bryant M (2016) *Command and Control* [online]. Available at: https://medium.com/elitecommandtraining/command-control-4854b2ffbd65 (accessed February 2019).

Butler PC, Honey RC & Cohen-Hatton SR (2019) Development of a behavioural marker system for incident command in the UK Fire and Rescue Service: THINCS. *Cognition, Technology and Work*. https://doi.org/10.1007/s10111-019-00539-6

Cabinet Office (2004) *The Civil Contingencies Act: Preparation and planning for emergencies: responsibilities of responder agencies and others* [online]. Available at: https://www.gov.uk/preparation-and-planning-for-emergencies-responsibilities-of-responder-agencies-and-others (accessed February 2019).

DCLG (2009) *Integrated Personal Development System – Code of Practice*. DCLG: London.

DCLG (2012) *Fire and Rescue Service Operational guidance: Operational risk information* [online]. Available at: https://www.gov.uk/government/publications/operational-guidance-for-the-fire-and-rescue-authorities-operational-risk (accessed February 2019).

Department for Communities and Local Government (DCLG) (2013) *Fire and Rescue Authorities: Health Safety and welfare framework for the operational environment* [online]. Available at: https://www.gov.uk/government/publications/health-safety-and-welfare-framework-for-the-operational-environment (accessed February 2019).

Dennett M (2004) *Fire Attack: An integrated strategy*. Bury: Summerseat Publishing.

Endsley MR (1995a) Toward a theory of situation awareness in dynamic systems. *Human Factors* **37** (1) 32–64.

Flin R & Arbuthnot K (Eds) (2002) *Incident Command: Tales from the hot seat*. London: Routledge.

Flin R, O' Conor P & Crighton M (2008) *Safety at the Sharp End: A guide to non-technical skills*. Boca Raton, FL: CRC Press.

Gough W (2012) *The STAR Model, Personal Safety for Firefighters at Incidents*. Birmingham: WMFS.

Hawley C (2004) *Hazardous Materials Incidents* (2nd Edn). New York: Thomson.

Health and Safety Executive (2010) *Striking the Balance Between Operational and Health and Safety Duties in the Fire and Rescue Service* [online]. Available at: http://www.hse.gov.uk/services/fire/duties.pdf (accessed February 2019).

Health and Safety Executive (2011) *Heroism in the Fire and Rescue Service* [online]. Available at: www.hse.gov.uk/services/fire/heroism.htm (accessed February 2019).

HM Govt (2008) *Fire and Rescue Manual, Volume 2, Fire Service Operations, Incident Command, 3rd Edition*. London: TSO.

HM Government (2013) *Emergency Response and Recovery: Non statutory guidance accompanying the Civil Contingencies Act 2004* [online]. Available at: https://assets.publishing.service.gov.uk/government/uploads/system/uploads/attachment_data/file/253488/Emergency_Response_and_Recovery_5th_edition_October_2013.pdf (accessed February 2019).

Hammond JS, Keeney RL & Raiffa H (1998) The Hidden Traps in Decision Making, Boston, Harvard Business Review. Available at: https://hbr.org/1998/09/the-hidden-traps-in-decision-making-2 (accessed February 2019).

Health and Safety Executive (2010) *Striking the Balance Between Operational and Health and Safety Duties in the Fire and Rescue Service* [online]. Available at: http://www.hse.gov.uk/services/fire/duties.pdf (accessed February 2019).

Hillson D & Murray-Webster R (2017) *Understanding and managing risk attitude*. Aldershot, UK: Gower Press.

Home Office (1981) *Manual of Firemanship Book 11 and 12, Practical Firemanship I and II*, London: HMSO.

Honeycombe G (1976) *Red Watch*. Jeremy Mills Publishing.

Klein GA (2008) Naturalistic decision making. *Human Factors: The Journal of the Human Factors and Ergonomics Society* **50** (3) 456–460.

Metropolitan Police (2019) *Emergency Boarding Up* [online]. Available at: https://www.met.police.uk/advice/advice-and-information/bu/emergency-boarding-up/ (accessed February 2019).

National Operational Guidance Programme (2018a) *The Foundation for Breathing Apparatus* [online]. Available at: https://www.ukfrs.com/foundation-knowledge/foundation-breathing-apparatus (accessed February 2019).

National Operational Guidance Programme (2018b) *The Foundation for Incident Command* [online]. Available at: https://www.ukfrs.com/foundation-knowledge/foundation-incident-command (accessed February 2019).

Noll GG, Hildebrand MS & Yvorra J (2005) *Hazardous Materials: Managing the incident*. Chester, MD: Red Hat.

Norman J (2012) *Fire Officer's Handbook of Fire Tactics*. Tulsa: Penwell Publishing.

Skills for Justice National Occupational Standard (NOS) (2003) *EFSM2 - Lead, Monitor and Support people to resolve operational incidents*. Available at: http://www.skillsforjustice-ipds.com/nos (accessed February 2019).

Sinclair M (2005) Intuition: myth or a decision-making tool? *Management Learning* **36** (3) 353–370.

Toft B & Reynolds S (1997) *Learning from Disasters: A management approach* (2nd Edn). Leicester: Perpetuity.

Watts BD (2004) *Clausewitzian Friction and Future War* [online]. Available at: https://www.clausewitz.com/readings/Watts-Friction3.pdf (accessed February 2019).

Legislation

HM Govt (1974) *Health and Safety at Work Act (1974)*

HM Govt (1998) *The Human Rights Act (1998)*

HM Govt (2000) *Management of Health and Safety at Work Regulations (1999)*

HM Govt (2004) *Civil Contingencies Act (2004)*

HM Govt (2004) *Environmental Information Regulations (2004)*

HM Govt (2004) *Fire and Rescue Service Act (2004)*

HM Govt (2005) *Regulatory Reform (Fire Safety) Order (2005)*

HM Govt (2007) *Fire and Rescue Services (Emergencies) (England) Order (2007)*

HM Govt (2007) *Corporate Manslaughter and Corporate Homicide Act (2007)*

HM Govt (2010) *Environmental Permitting (England and Wales) Regulations (2010)*

HM Govt (2015) *The Environmental Damage (Prevention and Remediation) (England) Regulations (2015)*

HM Govt (2015) *The Control of Major Accident Hazards Regulations (2015)*

Websites

Institution of Fire Engineers: Firefighter Safety – https://www.ife.org.uk/Firefighter-Safety

UK National Operations Guidance – https://ukfrs.com

Appendix A: Tactical plans: some examples

The following generic examples of tactical plans could be used as rough guidance for resolving three of the most common incidents types – structural fires, road traffic collisions and hazardous materials incidents. These plans follows the generic incident management process as outlined in the decision control process (DCP):

1. **Situation assessment:** information gathering and risk assessment (incident, resources, risk and benefits).

2. **Decision making and planning:** objectives, tactical priorities and operational tactics:

    a. Determining priorities and setting out a strategy.

    b. Determining the tactics required to satisfy the strategic objectives.

    c. Assessing resource requirements.

3. **Action:** implementing the tactical response:

    a. Setting up communications network.

    b. Organising the incident ground – sectors, responsibilities, management structure, supporting structures, allocating resources.

4. **Decision controls:** reviewing progress against milestones (expectations) while monitoring the situation continuously. Ensuring that risks are minimised and likely benefits are optimised at all times. Ensuring that the strategic objectives are still valid.

The extent to which an IC will follow the decision control process will depend on their skills, knowledge, experience and familiarity with the type of situation faced. The incidents below are based on actual real life situations. Comparing the sequence of the events with the DCP process is a useful exercise that shows that the evolution of an incident is not always sequential or linear. Sometimes actions are undertaken 'out of order', but they always return to the DCP processes (or other decision support tools systems).

Tactical plans

Incident 1: structure fires

Figure A1 1: fire in multiple occupancy building

Controlling the incident ground

Immediately upon arrival at incident, the IC should be seeking to determine the areas of greatest risk and consider where to put cordons in order to protect members of the public. This will also enable firefighters to carry out their tasks with a minimum of interference and disruption. It may not always be possible to do this due to higher priority activities such as carrying out rescues or preventing rapid escalation of the fire. Using a fire engine or other vehicle to block a street is a quick and effective method of restricting vehicle access and keeping members of the public away from the scene of operations. Plastic hazard tape or wooden trestles as barriers take time to put in place but should be considered as this will release firefighters for other tasks. When police officers are in attendance, they may be used for controlling barriers and monitoring access.

Rescues

Carrying out rescues of persons trapped in buildings on fire is the highest priority and should be carried out with a minimum of delay. Dependingt upon the number of firefighters and equipment carried on the first attending vehicles, rescues can be attempted immediately with only the minimum necessary safety measures in place if the risk to the victim is so great that serious injury or death may occur without that intervention. Circumstances in which immediate deployment may be necessary could include a casualty seen at a window from which smoke is emitting, or a collapsed person in the doorway of the building on fire. Where the presence of a casualty is uncertain (e.g. where there may have been squatters present) or a fire is so well-developed that the risk to firefighters is extreme, then a more controlled approach is needed so that safe systems of work can be fully implemented. The speed of intervention must balance the additional risk to firefighters with the likelihood of success of any rescue operations.

Locating the key areas of the incident

Having carried out any rescues, the next task is to look at the key areas of an incident. A reconnaissance (or '360°') enables the IC to review the overall incident scene where possible and obtain an appreciation of the scope of the incident and hopefully identify locations where the risk of spread is greatest and where blocking jets to prevent the spread of fire may be required.

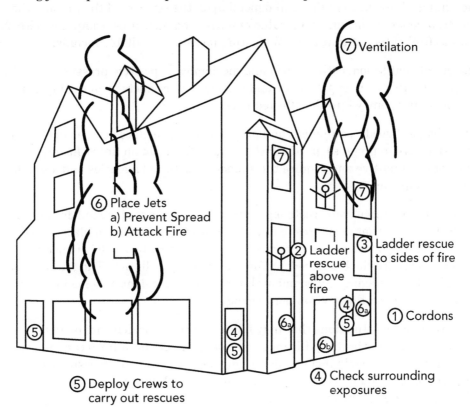

Figure A.1.2: sequential actions at a building fire

Containing, surrounding and extinguishing the fire

Attacking a fire without considering the potential impact may unwittingly casue the fire to spread. Understanding fire dynamics and the layout of the building will allow the IC to select the most appropriate tactics. Having located the fire, jets should then be put in place to block its spread. For example, attacking a front row fire from outside of a building may cause air to become entrained and 'push' the fire into the untouched remainder of the premises. Where 'blocking' jets are positioned first, the fire becomes contained and reduces spread to unaffected parts of the building. The use of positive pressure ventilation (PPV) techniques has also been used.

Once the spread of the fire has been checked, additional jets can be deployed to surround the fire. Extinction then becomes a matter of having sufficient water applied to the incident.

Mitigating the impact of the incident and supporting recovery

Activities to reduce the impact of a fire can start at any point during an incident: the timing of any intervention will depend upon the severity of the fire and the potential benefits (including the value of mitigation measures) compared with the risks to firefighters at that time. Mitigating measures available include:

The use of ventilation, both natural and mechanical – positive pressure ventilation (PPV) – to expel smoke and combustion products from the building and prevent its entry into the affected parts.

The salvage of goods, equipment and property can take place even as firefighting operations are in progress, particularly where a fire is on an upper floor. This may be resource intensive and require the support of external agencies and specialist salvage companies.

Contaminated run-off can create a pollution hazard and it is incumbent upon the IC to ensure that measures to limit the risk of this are put in place as soon as possible. The development of environmental strategy by the IC or specialist officer will reduce the likelihood and/or extent of any pollution. The failure to set up a plan for managing environmental impact of an incident may lead to future prosecution by and the environment agency.

Preparing for investigations that may follow a serious incident (involving a fatality, large losses or a significant operational success or failure) can aid investigators. This could include identifying key witnesses, authorising the preparation of statements by firefighters and controlling access to the premises to avoid disturbing potential evidence.

Incident 2: road traffic collisions

The process of managing a road traffic collision broadly follows the same pattern as that for managing fires, but will necessarily require a greater degree of co-operation and collaboration with other emergency services and agencies, including highway agencies and local authorities.

Figure A1 3: note all services present and an air ambulance landing at rear

Controlling the incident ground

On roads that are still running, the risk to crew safety is elevated and every effort should be made to stop the flow of vehicles immediately. Where the police service is already in attendance, this may already have been completed but were not then an RS vehicle should be used to provide a barrier to ensure a safe working environment. In order to maintain this safe zone and organise it such that either FRS vehicles and ambulances can gain access and egress, a joint service meeting should be held as soon as possible. A joint strategy for ensuring site safety, access and preservation of evidence should be agreed as soon as possible with all relevant agencies.

Locating the key areas of the incident

Having secured the site, the IC should make the scene assessment, identifying and prioritising casualties who are trapped and/or injured, and identifying risks such as fuel spillages or the contents of the vehicles themselves. While many road traffic collisions have a relatively small operational footprint, the IC should remember that the ejection of occupants of vehicles may occur at an impact point and casualties may be some distance from the resting place of the vehicles. A full examination of the scene should therefore be carried out as soon as possible.

Containing the incident and preventing casualties from deteriorating

A key priority at a road traffic collision is to carry out a casualty assessment and begin vehicle stabilisation procedures. Paramedics and medically trained firefighters can begin stabilising casualties immediately, before to extraction/evacuation. Working closely with medics, firefighters may then begin the process of extracting casualties.

Mitigating the impact of the incident and supporting recovery

Although the control of the incident remains with the police service, the FRS will be required to neutralise risks associated with the spillage of chemicals and fuels at incidents and provide appropriate support to other agencies which may be involved in recovery and restoration of the highway. Again, it may be necessary for the production of an environmental strategy plan to ensure any pollution produced will be limited and contained as far as is practical.

Incident 3: hazardous materials spillage

Unlike many types of incidents, firefighters attending hazardous materials often have the luxury of time in which to manage and control the incident effectively. A systematic process to manage the event is nevertheless required to ensure the safety of the crew and minimise the impact on the environment.

Figure A3 4: Spillage of pesticides adjacent to river (Exercise)

Controlling the incident ground

Having approached the incident using the most appropriate route (ideally from up wind and uphill), it is important that first responders set up an outer cordon. This outer cordon denotes the interface between the 'cold' and 'warm' zones and balances the need for safety of firefighters and the public and keeping the incident ground a manageable size. The organisation of the incident ground into hot, warm and cold zones will be determined by the IC following a reconnaissance of the incident, the topography and meteorological conditions, and in conjunction with specialist advice.

The cordons are controlled as follows: the police service manage the outer cordon surrounding the warm zone and the FRS is responsible for management of the inner cordon separating the 'hot' and 'warm' zones. Setting up a decontamination point – emergency or full systems – at the upwind interface of the hot and warm zones will be decided by specialist staff working with the IC.

Locating the key areas of the incident

As part of the initial scene assessment, the ICA should attempt to identify key areas of the incident including:

- the source of any leakage or discharge of liquids or gases

- casualties who may or may not still be present in the initial 'hot' zone

- potential evacuation areas, particularly downwind and the safe areas to which casualties, evacuees and firefighters can remain in safety.

Containing the incident

While addressing any issues relating to casualties, including decontamination, activities will commenced to contain the incident either through stopping the leakage at the source, capturing the lost material through the use of dams, barrels or other vessels, or by blocking drains and other outlets (gullies or streams etc) in order to hold hazardous materials within natural ground depressions. In some high-risk sites interceptors or other drainage mechanisms might be used to capture any hazardous materials run-off.

Mitigating the impact of the incident and supporting recovery

Once casualties have been removed and decontaminated where necessary, and any leakage/spillage controlled and contained, the collection and disposal of spilled materials need to be considered. This is now the non-emergency phase of the incident and, depending upon location and jurisdiction, it is likely that the collection and disposal of materials will be the responsibility of the organisation that caused the pollution hazard. The FRS is unlikely to be directly involved in this process but will be available to give specialist advice regarding the nature risks and hazards. Where there may be an impact on the environment and water courses as a result the spillage, it may be necessary for the FRS to support other agencies in reducing the damage.

At the end of an incident, contaminated personal protective equipment (PPE) and hoses and containment systems need to be cleaned or disposed of. Staff, if exposed to hazardous materials, will need to be managed, treated if necessary, and details of any exposure/contamination recorded in accordance with health, safety and welfare legal requirements.

Finally, as with all incidents, the lessons learned should be captured and disseminated in order for operational improvements to be made.

Appendix B: Identifying stress in colleagues

Behavioural indicators

Impact: A change in a normal patterns of behaviour can indicate acute stress.

Signs and symptoms:

- Hyperactivity
- Irritability
- Aggression
- Becoming withdrawn or detached from situations
- Apathy

Emotional indicators

Impact: Experiencing stress can lead to emotional distress.

Signs and symptoms:

- Vulnerability or a feeling of loss of control
- Panic
- Anxiety
- Fear of situation
- Fear of failure

Physical indicators

Impact: Situations that feel stressful can result in the body releasing adrenaline. This can result in some physical changes that can indicate stress.

Signs and symptoms:

- Increased heart rate
- Pupil dilation
- Sweating
- Dry mouth
- Butterflies in stomach
- Trembling

Cognitive (thought process) indicators

Impact: In manageable amounts, some stress may enhance cognition through clearer thinking and improved memory.

Signs and symptoms:

- Disrupted concentration
- Difficulty prioritising objectives or tasks
- Narrow focus or tunnel vision
- Lack of focus on relevant cues
- Memory impairment information overload leading to forgetting
- Task overload due to the limited capacity to undertake tasks resulting in tasks being dropped or not completed
- Easily distracted
- Focuses on information that supports mental picture of the situation, while ignoring other pieces of information
- Forgetting to carry out an intended action at the time it was due to be done
- Impaired decision making
- Blank mind
- Resorting to familiar or drilled routines

Appendix C: Enviroment Agency Risk Assessment form

National environmental risk assessment

National Operational Guidance Programme

Date		Time	
FRS Incident No.			
EA NIRS No.			

Details of persons completing this assessment

FRS	

Incident address/location

Polluter information

EA	

General overview

| Is there a risk to life or health? | Yes ☐ No ☐ | Have suitable environmental control measures been deployed? | Yes ☐ No ☐ | Has the appropriate environment agency been informed? | Yes ☐ No ☐ | Is the EA in attendance? | Yes ☐ No ☐ |

Information available from the environment agencies:

EA hostile site or illegal site*	Industrial site* COMAH
Hydrology* (observation)	Industrial site - permit
Monitoring * (flow)	Emergency plan/site specific risk information**

* Information from EA mapping
**Information from EA database

Step 1 - Gather risk information

Source (environmental hazards)	Pathways	Receptors
Fire water run-off ☐	Surface water (rivers, ditches etc.) ☐	Nearby population ☐
Smoke plume ☐	Drains/utilities (surface and foul sewerage systems) ☐	Surface water (rivers, ponds, lakes, coastal waters etc.) ☐
Physical damage (trompling, compression, parking) ☐	Vehicles/transport ☐	Areas of natural conservation ☐
Gases ☐	Wind/airborne ☐	Environmental protection designations (source protection zones, groundwater, etc.) ☐
Fuels/oils (mineral / vegetable) ☐	Groundwater/permeable soil ☐	
Organic substances (e.g. milk, beer, fruit juice, sewage, etc.) ☐	Others Please specify	Drinking water (aquifers, reservoirs, etc.) Permits ☐
Chemicals ☐		Fisheries, bathing waters ☐
Pesticides/fertilisers ☐		
Foams/CAFs ☐		Others Please specify
Detergents ☐		
Radioactive hazards ☐		
Biohazards (e.g. blood, bacteria, etc.) ☐		
Hazardous fly-tipped waste ☐		
Suspended solids (e.g. sand, silt, etc.) ☐		
Utilities (gas/oil pipelines) ☐		
Others Specify		

If there are entries in all three elements from this step it is likely that a risk to the environment is present.

Source ✚ Pathway ✚ Receptor = Environmental Risk

Action should be taken to **Eliminate, Reduce, Isolate** or **Control** the environmental hazard.

Step 2 - Draw a plan or describe the incident (Include location of hazards, pathways and receptors:)

Step 3 - Impact

Estimate the likely impact of any environmental damage?

☐ **Low** — *It is unlikely that environmental damage will occur. Take reasonable action to reduce the risk.*

☐ **Medium** — *There is significant evidence to suggest that environmental damage may occur. Take action to protect the environment and monitor.*

☐ **High** — *The receptor is located within an area of high environmental risk. It is likely that environmental damage will occur.*
Prioritise and action to contain pollutants in liaison with the relevant environmental agency.

Step 4 - Actions

Control measures		Environmental protection actions in place	
Liaise with relevant EA/conservation body	☐	1 Contain at source	*Details of action taken (include equipment used)*
Liaise with hazmat adviser (HMA)	☐		
Contain / absorb (e.g. Grab pack EPU equipment, etc.)	☐		
Reduce quantity of media being applied	☐	2 Contain close to source	*Details of action taken (include equipment used)*
Controlled burn	☐		
Alternate firefighting media	☐		
Recycle firefighting water	☐		
Dilution	☐	3 Contain on the surface	*Details of action taken (include equipment used)*
Discharge to foul sewer (with approval from sewerage undertaker)	☐		
Extinguish fire	☐		
Accelerated controlled burn	☐	4 Contain in drainage system	*Details of action taken (include equipment used)*
Air quality advice from relevant body	☐		
Removal/separation/quench in bund	☐		
Treatment/disposal	☐		
Defined access/egress route (protected areas)	☐	5 Contain on or in watercourse	*Details of action taken (include equipment used)*
Water spray to reduce plume	☐		
Others *Please specify*			

Supporting information

In addition to the environmental risk assessment and any subsequent action taken to protect the environment, the relevant environment agency should be informed where:

- The flow of water being pumped at an incident exceeds 4000 litres per minute *(approximately)*
- The incident involves radiation, hazardous materials or CBRNe
- Where decontamination operations are in use
- Where the substances are above normal domestic quantities or where it is confirmed that quantities of products involved reach the suggested thresholds below.

25 litres or more	
	Oil or fuel
	Firefighting foam concentrate
	Detergents including: *Washing powder, washing-up liquid, shampoos, soaps, car cleaning products, etc.*
	Disinfectants including: *household bleach, Dettol, etc.*
	Paints and dyes
	Cooking oils, glycerine, alcohols
	Cutting lube or water-soluble polymers

250 litres or more	
	Food products, particularly: *sauces, sugars, salt, syrups, milk, cream, yogurt and vinegar*
	Any beverage, including: *soft drinks, beers, wines and spirits*
	Organic liquids/solids, including: *blood, offal, farmyard slurries, sewerage sludge, anti-freeze*
	500Kg of sand, silt, cement, chalk, gypsum/plaster

BV - #0023 - 230724 - C22 - 246/186/15 - PB - 9781912755097 - Gloss Lamination